T0321917

Flexible Assembly Systems

Assembly by Robots and Computerized Integrated Systems

Flexible Assembly Systems

Assembly by Robots and
Computerized Integrated Systems

A. E. (Tony) Owen

Tony Owen, M.B.A., Ltd.
Bletchley, United Kingdom

Plenum Press • New York and London

Library of Congress Cataloging in Publication Data

Owen, A. E.
   Flexible assembly systems.

   Bibliography: p.
   Includes index.
   1. Assembly-line methods. 2. Robots, Industrial. 3. Automation. I. Title.
TS178.4.O94 1984   670.42'7   84-4852
ISBN 0-306-41527-5

Foreword

It has become clear in recent years from such major forums as the various international conferences on flexible manufacturing systems (FMSs) that the computer-controlled and -integrated "factory of the future" is now being considered as a commercially viable and technically achievable goal.

To date, most attention has been given to the design, development, and evaluation of flexible machining systems. Now, with the essential support of increasing numbers of industrial examples, the general concepts, technical requirements, and cost-effectiveness of responsive, computer-integrated, flexible machining systems are fast becoming established knowledge.

There is, of course, much still to be done in the development of modular computer hardware and software, and the scope for cost-effective developments in programming systems, workpiece handling, and quality control will ensure that continuing development will occur over the next decade.

However, international attention is now increasingly turning toward the flexible computer control of the assembly process as the next logical step in progressive factory automation. It is here at this very early stage that Tony Owen has bravely set out to encompass the future field of flexible assembly systems (FASs) in his own distinctive, wide-ranging style.

The structure of the book owes a great deal to a broad FAS study conducted by British Robotics Systems Limited, and this results in what may be described as a useful case-study orientation. This book is without doubt the first of its kind in its field, and usefully attempts to identify, define, and quantify many of the aspects, components, and implications of future FASs.

While the text (if not provocative) is thought provoking, it is clearly written to enable the nontechnical reader to understand the author's view. I am sure that impending industrial development of FASs will see considerably more written on this important subject in future years. However, I applaud the author for his initiative and courage in addressing such an important field now, in these, its early days of emergence.

KEITH RATHMILL
Robotics & Automation Group
Cranfield Institute of Technology
Bedford, England

Preface

The subject of this book is the flexible assembly system. It is discussed and evaluated here as a complete and separate entity from any other form of automated system within the manufacturing industry. For the purposes of continuity, this book draws upon, but is by no means limited to, a British government funded feasibility study into the viability of an advanced flexible assembly system for a "green field" site.

The purpose of this book is manifold in that it enables the findings of the feasibility study to be disseminated in a generalized manner, whereby they can be adapted and applied to many diverse assembly-related activities. This book also permits the presentation of up-to-date facts against the background of a real-world study, as well as defining both flexible assembly systems (FASs) and flexible manufacturing systems (FMSs) in terms that are both precise and unambiguous. Finally, the book hopefully fulfills the need for a definitive volume outlining the philosophies that lie behind flexible assembly systems.

The reader is guided through the subject in a logical and systematic manner. Thus, the first chapter deals with defining the tasks that are to be conducted within the FAS and establishing the structure in which they are to operate, as well as considering the implications for the commercial sector. Next, computerized material transfer is examined as being a prerequisite for a fully automated system. Chapter 4 deals with component presentation equipment and fixing and forming techniques, as well as the design of the work station. Design of the product for automated assembly is covered in Chapter 5, while Chapters 6 and 7 cover people, robots, and their respective and related benefits and problems. The next chapter discusses machine senses and artificial intelligence, and is followed by Chapter 9 which covers failure analysis as a quantitative subject, including the determination of manning levels and the implication of humans on the operating efficiency of the FAS. The final chapters look in depth at the economics of the FAS and, gazing into a crystal, forecast the FAS as one of the precursors of the factory of the future.

Contents

Abbreviations

AS/SR   Automated storage/retrieval systems
BRA   British Robot Association
BRSL   British Robotic Systems Limited
CAD   Computer aided design
CAF   Computer automated factory
CAM   Computer aided manufacture
CNC   Computer numerical control
CPU   Central processing unit
DNC   Direct numerical control
EOQ   Economic order quantity
EPROM   Eraseable and programmable read only memory
FAS   Flexible assembly system
FMS   Flexible manufacturing system
GT   Group technology
LED   Light emitting diode
KS   Knowledge source
MRP   Material requirements planning
MTBF   Mean time before failure
MTTF   Mean time to failure
MTTR   Mean time to recovery (repair)
NC   Numerical control
OCR   Optical character recognition
OLRT   On-line real time
PCB   Printed circuit board
RIA   Robot Institute of America
VDU   Visual display unit
U.K.   United Kingdom
U.S.A.   United States of America
WIP   Work in progress

Introduction

In 1980, the British government funded a feasibility study into the viability of a flexible assembly system (FAS). The study, which was conducted by British Robotic Systems Ltd. (BRSL) of London, took eight man-years to complete and resulted in a 15-volume final report that had a restricted circulation.

The study's aim was to analyze the parameters and viability of an advanced flexible assembly system that would permit the assembly of products defined only in the vaguest functional outlines. The facility would also have to incorporate and be amenable to the latest assembly techniques and philosophies.

In productivity terms, the proposed plant would have to have a value, added per head, some three times greater than that of a conventional plant of the same size. Unfortunately, while the feasibility study cost $300,000 it has not (yet) directly resulted in a new factory being built. The primary reasons are the recent inflated rate of money, the depressed market, and the general socio-economic state of the United Kingdom.

It is important to distinguish between the FAS and the flexible manufacturing systems (FMSs), the latter of which have been the source of a lot of books, papers, reports, and speculation within the manufacturing industry fraternity. These systems (FMSs) were first conceived in the 1960s by Theo Williamson with his System 24. For many reasons, including the lack of compatible computer systems in a suitable price–power range, they spontaneously aborted. Nowadays the FMS is a viable reality that represents a natural progression, via numerical control (NC) and direct numerical control (DNC) machining centers, from the traditional job shop of a previous age. There are presently some 100–125 FMSs in operation or about to be commissioned throughout the world.[1]

FMSs in the right environment are used to perform job shop or batch production machining, i.e., they can process a single component or several thousand components with the same increase of efficiency over traditional methods. Figure 1 shows the low productivity that can be expected with traditional systems because of the time wasted in nonproductive activities, such as "between operations" storage, tool setting, etc. Additionally, long lead times are associated with these archaic methods.

Therefore, the primary use of a FMS is the removal or shaping of metal to produce a component. Hence, the important aspects are those of processing feeds and speeds, tooling lifetimes and automatic monitoring, swarf removal and separation, and the reclamation of valuable oils and metals from the process for recycling.

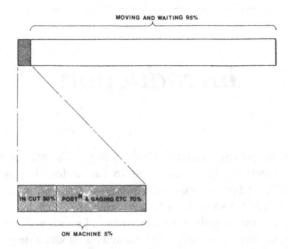

MOVING AND WAITING 95%

IN CUT 30% POST^N & GAGING ETC 70%

ON MACHINE 5%

Figure 1. The throughput time of a given product is mainly absorbed by the nonproductive activities of moving and waiting. These are very prevalent within the traditional manufacturing industry where, as shown in the figure, only 1.50% of the throughput time is spent in adding value to the product.

On the other hand, the function of a FAS is the processing of components and materials into a product. Here the emphasis is on the compatibility of the components for high assembly rates, automatic testing for functional quality, and the conservation of consumable materials. It does not include any shaping or forming of materials except where the process is an essential and integral part of an assembly task.

The two systems are in fact intertwined and make use of the same philosophies and technologies. They are both constituent parts of, and precursors of, the anticipated computer automated factory (CAF) of the 1990s, whereby raw material will enter a facility in which it will be converted into components for use within specific products that are being made to order by the CAF. Both FASs and FMSs use the philosophy of quantification of system elements, which results in consistent quality, predictable output, and maximized system capacity. Also, they both use robotics, remote conveyor systems, and computerized storage systems.

The fundamental differences are shown in the following two definitions:

(i) An *FMS* is a computer controlled automated machining system for converting raw material into *components* of a known and desired *geometric* quality.

(ii) An *FAS* is a computer controlled automated assembly system for converting raw material and purchased components into *products* of a known and desired *functional* quality.

Definitions apart, the manufacturing industry of the late 20th century must be of a dynamic nature so that it can respond quickly and correctly to a change in demand for its products from the market place. The major activity within the manufacturing industry is that of *assembly,* which accounts for 53% of the time and 22% of the total labor incurred in the manufacture of a typical product.

Unless a manufacturer is in a monopolistic industry, the demand for its goods varies with the capricious nature of the consumer, the number of functional competitors (not necessarily quality competitors), alternatives available, and the quality fluctuations of the particular item. If the manufacturer is to survive, then it should have a production process that can be varied to suit a range of products so that *product specific* output can be correlated to *product specific* demand patterns.

It follows that a production system that is flexible and allows for control of all aspects of the manufacturing process in general, and assembly tasks in particular, offers the potential for high productivity, with the resulting financial benefits to the company.

The traditional assembly philosophies have focused on processes and machines to be grouped in technological ghettos (not to be confused with group technology), without any real consideration for fast throughput of products. Reorganization of processes and machines into product cells means that identifiable logistic lines of manufacture and communication be both visible and definable.

When a process is known and defined, management can assess its efficiency in a quantitative manner, identify problem areas, and resolve or minimize the effects of the problem. This is vastly different from the "normal" manufacturing and assembly areas that have evolved in such a way that they can process "everything" that comes along. Thus system balancing cannot readily be achieved and neither can scheduling or output rate be accurately foretold.

The concept that is being promoted within this book is that of a FAS, whereby flexibility allows for any product within the functional, power, and geometric constraints of that assembly system to be processed. Since the "product" is defined the system can be designed such that it can process the products quickly and effectively without spending a lot of its "in process" time as work in progress (WIP) on the shop floor. Because we are dealing with a defined system envelope, it can be controlled to the extent that its constituent elements and boundaries are known and defined.

The major advantage of a FAS over hard automation is that its structure can be varied at will by the inclusion/exclusion of particular assembly modules such that throughput is maximized. The size of the FAS is obviously a function of the product parameters. If the existing and undefined products are grouped into families, then the complexity and/or possible interlinking of a number of sub or major FASs to satisfy a particular product group's requirements can be evaluated.

For any automated system the availability of the correct parts at the right place and right time is paramount to its success. Product variability means that the storage system must be able to respond quickly to changing requirements. Also, since the storage costs are high, it makes sense to maximize the storage density such that the costs of storage are minimized. High-density storage and the fast presentation of the correct parts leads to computer automated warehouses, whereby the computer is solely responsible for the identification, storage, and retrieval of all materials. The inclusion of humans within such a system would result in chaos, hence their presence is limited to that of overseer.

Consumers purchase goods on two criteria, those of price and quality. Irrespective of the price, it is important that the quality of a product be consistent since variable quality can only lead to scrapped products, overexpensive production costs, and a lot of consumer problems. The discussion here is not about high or low quality (since that is very subjective and cost related), but about the methods of manufacture such that the products are of a known level of quality that is compatible with the production processes and the desired factory cost of the product.

Quality is dependent upon the quality of the purchased materials and components *and* the assembly process, both of which can be checked by standard inspection techniques. The purchased goods should be checked as they are delivered, by whatever quality control technique is determined suitable. It should always be remembered that the "true value" of an item to the company may be vastly different than its cost. Directly associated with quality control is stock control, for which there are similar guidelines concerning the correct procedure for various valued parts, as well as the consequences of stock outs.

If a product is to be assembled efficiently, it must be processed in an environment that is conducive to high productivity. This means that the various components and subassemblies must be presented in the correct place, at the correct time, and in the correct attitude and orientation. The work station must therefore be designed such that all components can be accessed easily and correctly within the operational envelope of the assembly tool. If a person is involved, then the work zone is restricted to the limitations of human reach and flexibility. The use of a robot or assembly machine often means that the arrangement is more rigid and formalized. Work stations that are designed as modules for either robot or human usage will obviously have to be biased towards the human's needs and capabilities.

Fundamental to the efficient use of any automated assembly technique is the design of the product itself. Design for automated assembly means that the product is arranged such that its assembly can be achieved without contorted movements of the assembly or the assembler. Ambiguity of assembly must be avoided at all costs. This is usually achieved through the use of geometric redundancies such that the component has either a unique mode of assembly or it can be assembled without prejudice, no matter how it is presented.

The use of people as productive elements should be avoided since they introduce variability, inconsistency, and nonquantitative factors into any productivity equation. Where they should be used is for tasks that take advantage of the natural human attributes of intelligence, i.e., they should be used for the purposes of supervision, inspection, and design.

Removal of people from tasks that are boring, degrading, or performed in an environment that is not conducive to personal well-being, is essential by both law and social ethics. Any person that is employed in a human-oriented task will perform at a better productivity level than if employed in a subhuman task. The net result is that the company's gross output per person will be increased. The direct involvement of certain persons with automated systems in general and robotic systems in particular can lead to psycho-sociological problems as a direct result of the differences between human and machine work ethics.

No matter what form of assembly system is developed it must be compatible to quantitative assessment. This will take the form of technical output, up time, down time, reasons for failure—as well as the economic criteria of break-even points, pay-back periods, fixed and variable costs, etc. Operational efficiency must take into consideration each and every element within the system, such that the total system is represented. This means that the weak and/or defective areas can be pinpointed for investigation.

Specification of System

A FAS is structured to satisfy a series of tasks as specified by management. Prior to the analysis and specification of the operational systems within the FAS, it is necessary to state what is required from the FAS and how it will be structured so that control is assured.

Although each FAS consists of two primary sectors (conversion and commercial), consideration of how they interact is often ignored or forgotten when the overall efficiency of a FAS is being determined. This chapter examines task definition, structure, and the commercial sector, with the aim to show their importance to both the viability and operational well-being of the FAS.

2.1. Task Definition

Before a FAS can be designed it must be defined. The definition must relate to the technical, managerial, economic, and social aspects of the system. The degree of productivity of the FAS will depend upon the quantitative nature of the definition. The lower the quantitative factor, the lower the likelihood that the system will meet expectations. This relationship is indicated in Figure 2.

The technical definition will describe the functions of FASs in terms of sophistication, novelty, and integration. The parameters will be those of the processes used and the "products" created. The mix of "state-of-the-art" technology and regular techniques will be balanced by the degree of automation and the computer control structure. The technical aspects of material transfer will be compared against the practicalities of a truly flexible system.

The managerial definition will describe how the FAS can be examined to determine (in real time) its productivity and efficiency indices. It will also be necessary for a preferred geo-economic environment to be specified such that the FAS is situated within, or provided with, the appropriate infrastructure for its tasks. The management definition will also include market forecasts for the product categories, mixes, and speed of product changeover.

The economic definition will describe the FAS and the products in terms of fixed and variable costs, productivity levels, and manning levels. The total system will be specified in terms of overall cost expenditure, payback period, and value added per head.

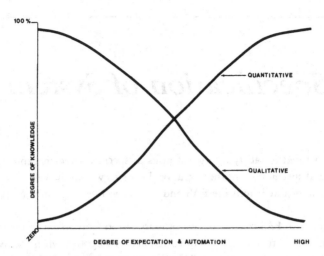

Figure 2. For any situation, the level of knowledge will be a mix of quantitative and qualitative information. As the ratio increases in favor of the quantitative, so does the realization of expectations of the system. While it is unlikely that 100% quantitative knowledge of any system can be achieved, it is essential that the target level should approach this ideal if the system efficiency is to be optimized.

The social definition will encompass the relationship between the FAS and its stakeholders. It will specify numbers and types of human employees, as well as the employee tasks. The subject of job enrichment will be linked to employee motivation as will the mix of human and nonhuman employees. The division of labor between the white, blue, and steel-collared workers will effect the social life of the locality, as will the pollution and effluent controls specified and imposed by state and federal codes and legislation. The ultimate definition under this category will be that specified under and inferred from the FAS's own code of social ethics.

A definition for a FAS could be as follows:[2]

> The FAS will realize a revolutionary concept, but it should not, in general, contain individual elements of unproven design. The concept is one of a small (2,800 square meter floor area) unit on two or three floors, employing approximately 250 persons. The unit should be capable of producing two to three times the added value per head that could be achieved in a conventional factory making the same products.
>
> The FAS will produce electronic/electro-mechanical devices of less than fifteen kilograms in weight, and generally less than 0.03 cubic meters in volume. The final product, in most cases, will be constructed from sub-assemblies, also made within the system, each typically constituting 10–30 percent of the final product size.
>
> The process technologies will be many and varied, and the FAS will need to be flexible enough to accept new ones with relative ease. It is important that the areas of technological growth for the next decade be identified, so that the FAS's fabric can be designed for easy incorporation.
>
> The intention is to produce 200,000 to 300,000 units per year with individual product types having volumes less than 100,000 and greater than 1,000. It will be necessary to run different products concurrently, and the minimum lead time from order to delivery will be between a week

and four weeks in general practice, assuming adequate documentation and availability of materials. It will also be possible to switch from one product to another in an unusually short time (one to two hours) thus combining the efficiencies of flow production with the flexibility of batch production.

The target is a FAS that will cost no more than a conventional factory, of the same output, i.e., 5,600 square meters and 500 persons. The intention is to achieve this target by developing a flexible integrated materials handling, storage, information and control system which feeds multi-product work areas. The work areas will be dedicated to specific processes and will have extensive computer assistance. Where feasible and appropriate the work areas will be robotic in nature, yet at the same time utilizing general purpose, or where necessary, dedicated automated equipment. The overriding priority must be given to the working conditions and the opportunity for job enrichment for the human employees.

2.2. Structure

If an FAS is to operate as a fully efficient and productive system, it is necessary that it be known on a quantitative basis in its elemental form. This means that the operation and performance down to the least significant part of the FAS can be predicted and that any performance variation can be detected and pinpointed before it becomes significant. Obviously, while 100% quantitative information on a given FAS is a target to be aimed for, the pragmatist will realize that the best that can be expected is a mix of quantitative and qualitative knowledge, with the ratio in favor of the former.

The ideal of FASs is that material will flow in an unidirectional high-speed motion, through a series of integrated work areas, with the minimum of nonproductive time spent during the transition. This is in fact the major difference between the FAS and the "normal" manufacturing company, where the various areas of responsibility act as independent and not interdependent entities. Also, the fact that machines are arranged in technological groupings leads to a lot of wasted time and effort, as the products are progressed from one group to another and from one machine to another within a given technological group. This can mean that the company is plagued with problems of inflexibility, high stocks, long lead times, and poor control of costs.

Within the FAS, it is the management that has the ultimate responsibility for efficiency and productivity, since it is they who through example and direction determine how it will function elementally and as an integrated whole. The amount of data generated within the FAS is vast, yet it is mostly relevant, necessary, and of prime importance to the FAS's decision makers. In order that decisions can be made, the information should be available when and where required and in a form that is relevant and understandable. It should be remembered that the 125-mm-thick fanfold of printout takes a lot of time to read and assimilate; consequently, it is not compatible with controlling a dynamic FAS in real time.

It is useful if the FAS can be neutralized so that its personality does not cloud the issues when strategic decisions have to be made. It is therefore necessary to segment the FAS such that the relationships between the various interdependent yet independent elements that constitute that *particular* FAS format, and their synergistic whole, can be related to the productivity level that is desired (or occurring). The structure can then be analyzed and, if required, modified to promote productivity, reduce costs, or improve and stimulate "on time" deliveries.

To provide this structure, it is necessary to have a database. The formal definition of a database is "a common pool of data, structured to provide timely and accurate information to all *authorized* users." The reason that the computer industry puts so much emphasis on the concept of how data is stored is not always apparent to "noncomputing" personnel. The primary difference between computer systems and manual systems is that the computer systems are fast and highly structured, and are consequently very difficult to change. It has become obvious over the past two decades that the best way to achieve a balance between conflicting requirements of flexibility and fast access is to separate the data and the programs by putting all of the data structure into a separate area again—between the two—giving three separate items. These are the actual data, the data structure and the access mechanisms, and the programs that do what the user wants (the applications programs).

The term database in fact really means the data and its description, whereas the term database management system means the method of structure and the programs which provide the access mechanism. In practice the term database is used loosely for both or either.

Database management systems are based on the concept of a "dictionary," defining the types of data and also the structures of the data in terms of its logical relationship to other pieces of data (part of the payroll record is hours worked) and also how it may be accessed. This may seem an impossible degree of formalization, but it is generally more flexible in practice. Prewritten database management systems are now commonly available. They vary in complexity and are very highly developed. Except where high speed or very specialized systems are required, the normal solution is satisfied by a standard database package. Price is in direct proportion to the complexity of the system and varies from $1,000 to $50,000.

At the database level all manufacturing companies, and consequetly all FASs, look pretty much the same, the basic variations occurring because of the different reporting requirements. Using database technology any system is divided into three fundamental pieces (rather than "*n*" pieces in a noncomputer system), one for each computer application.

The first piece is called the database and has already been explained. The second critical piece is what is called the *input control system,* which is generally geared to the organizational structure and is hence independent of the database. The third piece of the system structure is the *outgoing control system,* which is not similar to either of the other two pieces. This is because the output demands are always changing while the database and input control system are relatively stable. Hence the output control system must be designed as a dynamic system and responsive to all sorts of inquiries. Its flexibility is critical to the survivability of the whole automated system structure.

Data is created and used in one or more of four basic structural levels. The first level is that of *material flow,* which refers to the transformation of raw materials into finished goods. It is the only level at which physical movements of goods takes place, either in the form of the production of goods or the receipt of bought out items, and it requires a great mass of detailed information.

The second level is that of *process control,* which is concerned with supervision, quality control, and process unit control. As a consequence, only a small amount of information is required and, in general, the time duration for which the information is kept after creation is short. Only in the case of abnormal operating conditions is it necessary to store information for future analysis. *Production control* is the third level and is concerned with the hour-to-hour management and optimal utilization of the production facilities. It is therefore interested in scheduling, cost control, inventory control, and maintenance. It requires information of lesser detail than that of the process control, but needs to store significant information on the database.

The fourth level is that of *management control.* It has two sublevels, those of *strategy* and *tactics.* The first of these sublevels is concerned with the overall policy and requires brief summaries of past operations and information relating to external factors that influence the management of the company. For instance, market demand forecasts, current economic conditions, competitors, raw material availability, costs and social pressures. The second sublevel is concerned with putting the policy into practice, i.e., making production plans meet demand and making efficient use of resources. Therefore, more detailed information is required regarding internal operations than at the process control level.

It should now be clear that the overall efficiency of a FAS is fundamental to its profitability and ultimately to its survival. As shown in Figure 3, efficiency is related to productivity, which in turn is related to intimate knowledge of what is happening when and where, such that inefficient sectors, personnel, processes, or shifts, etc., are highlighted for management to act upon. Since a FAS is a dynamic entity, it is necessary that any reporting/analysis/feedback system is equally dynamic so that the reaction to a given situation is speedy. The development of microelectronics has enabled revolutionary changes to take place in all of these directions within industry. The low cost of microcomputers permits their introduction throughout all levels of the FAS, from the lowliest function on up through to the boardroom. The proliferation of on-line real time (OLRT) terminals at the lowest, yet most crucial level is significant to the productivity of the "conversion sector," since it is here that the raw materials are physically transformed into value added goods (or scrap). It is also here that profits are made and lost a cent at a time through the minute-by-minute decisions that are taken throughout the year, yet historically the amount of information available at this level has been limited, often incorrect, and out of date.

Microelectronics also permit data to be collected and assessed at different ter-

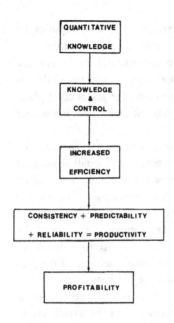

Figure 3. Profitability is the bottom line of any FAS. The level of profit is a function of the productivity of the system, which is itself directly related to the level of quantitative information, as well as the application of monitoring systems so that the system exhibits acceptable levels of consistency, predictability, and reliability for the product being assembled.

minals within the network, with access to various data files being regulated by appropriate passwords. This ensures that while the terminals at the lowest level will have limited, albeit sufficient, accessibility to information, they will covertly (and overtly) collect a mass of information for the higher "status" terminals. The highest-level terminals will have access to all of the available data and will collate, analyze, and present information in any desired format in real time, or as a batch mode process from a mainframe computer.

BRSL[2] showed that a three-tier hierarchical design would best suit their requirements. At the top of the "pyramid" will be the central mainframe computer performing tasks over long time periods. In the middle layer will be the midi- and minicomputers that control the FAS, and at the lowest level will be the microcomputers that react in milliseconds and operate in a restricted environment, controlling, for example, the joints of a robot in one of the work stations.

The central computer has two major functions. The first is to run all of the managerial functions such as personnel, accounting, sales, and planning. The second function is the retention and manipulation of the major system database to support the FAS's environment. These, therefore, are the long-time, large-number crunching routines, wherein accuracy is paramount to speed. The functions controlled by the central computer (shown in Figure 4) are:

(a) Inventory control;
(b) Sales analysis;
(c) Finished goods control;

(d) Automatic invoicing;
(e) Management reporting;
(f) Payroll;
(g) Personnel;
(h) General ledger;
(i) Product costing;
(j) Product definition and amalgamation;
(k) Engineering simulation;
(l) Computer-aided design (CAD);
(m) Computer-aided manufacture (CAM)—in terms of preparing programs for on-line loading of robots or machines.

The FAS control computer has as its main function the following:

(a) Storage of global data;
(b) Control and coordination of the next level of computers;
(c) Interfacing with the central computer;
(d) Maintaining the production schedule;
(e) Managing the automatic materials movement;
(f) Routing information between systems;
(g) Reporting and responding to inquiries; and
(h) Overall analysis of data.

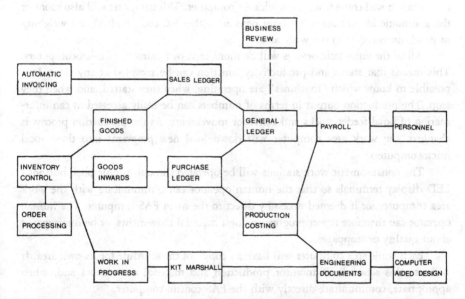

Figure 4. These functions and associations are directly controlled by the central (mainframe) computer. They invariably require a lot of number crunching with accuracy being more important than speed.

The FAS control computer will use a subset of the main database for operational purposes. This subset will be transmitted nightly from the central computer and updated during operation. Communication between the two computers will be continuous and reflect the dynamic operation of the FAS. The control computer also interfaces with the warehouse computer and the computers of the individual work areas (see Figure 5).

The warehouse control system will be largely independent and will operate by translating commands to move particular material between the warehouse and the output points into actual bin numbers and coded locations. To this end it will have a local file with information on the various materials and their locations, including material that has been issued to work areas, or is on conveyors, etc. Actual control of the cranes, conveyor timing, routing, etc., will be provided by the warehouse manufacturer's own supplied software. It is in effect a three-dimensional point-to-point system that moves bins from one location to another.

Each work area will be a separate, semiautonomous system. Each work area will be, at least in terms of software, identical to every other work area. A minicomputer will be situated and configured to suit each work area, so that "it" knows how many work stations there are in "its" area, whether they are manual or robotic, and the number of other automatic machines or processes that there are under its control.

Each work area computer will be responsible for the movement of materials to and from the top of the lifts and conveyors that move it (the material) between that work area and other locations within the FAS. Most of the detailed timing and control will be done by microcomputers attached to the conveyor systems, albeit that overall coordination will come from the work area computer. This computer will also monitor the automatic identification (by light pens and other bar code readers) and weighing of goods on entrance to the work area.

All of the automatic devices will be monitored by a number of microcomputers. This means that status and productivity function can be sampled at any time. It is possible to know which "machines" are operating, when they started, and when they stop. The production output in terms of numbers can be easily accessed as can information of bowl feeder stocks and conveyor movements. As the production process is changed, the work area computer will down load new programs into these local microcomputers.

The nonautomatic work stations will be operated through simple push buttons/ LED display terminals so that the human operator can communicate with the work area computer, or if deemed necessary direct to the main FAS computer. The human operator can therefore report progress, request material movements, or be interrogated about quality or stoppages.

Each work area computer will have a "copy" of the schedule for its own area. It will use this schedule to monitor production and material movements and, when appropriate, communicate directly with the FAS control computer.

Although the central computer may appear to have more to do, the FAS and subsidary computers will be doing far more repetitive and critical functions and should therefore require more power, if not more software. In other words, they require larger and faster computers to control their environment. An even more difficult problem is that of security. It is essential for the FAS systems to be secure, i.e., the computer should be available 100% of the time. Because the central and FAS computers overlap in one important area—that of inventory control—it is not realistic to regard the distributed network as 100% secure. Therefore, the FAS computer will have to maintain its own database for operational purposes, which means strict control of transmissions and frequent cross checks.

To obtain total security, every point of the system which has to be secure must be duplicated. In practice this means that the central processing unit (CPU) and associated peripherals must be duplicated, plus any critical terminals or automatic machines. It is important to realize that the cost of security is not just hardware, but also software, in order to ensure that the twin systems are "in step," which means a large programming effort. In addition, there is the logging of transactions, frequent copying of data, and so on.

An alternative to 100% security is fragmentation of the system through distributed processing, and estimation of the security required at each point. The configuration for distributed processing is given in Figure 5, with each work area having its

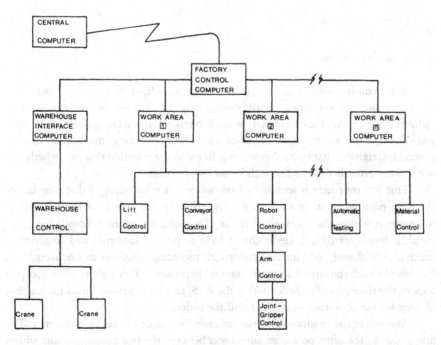

Figure 5. The FAS is controlled by a hierarchy of distributed computers, from the central computer through the factory control and work area computers to the micro's that control the robot appendages. With the exception of the central computer, all systems operate in on-line real time so that the dynamic operation of the FAS is achieved.

own local processor capable of independent operation. Failure of a local processor would halt local automatic processing and restrict manual processing to what could be achieved with materials currently on hand. As each local processor would be identical, spare modules would be kept and the mean time to recovery (MTTR) would be very low.

To have 100% back-up on each work area seems excessive. Industrial mini- and microcomputers are becoming increasingly robust and hardware failures should be no more likely than on other machines. Software failure is less likely on a single system and the program could be restarted from the central computer.

Failure of the central processor would halt all management functions and possibly, though not absolutely necessarily, material movements. The implications would be that each work station could keep on working, probably at a slower rate because the local material transfer system would have to be operated by hand.

All of the above is not to suggest that a second computer would be needed, only that the second unit could be on "cold" standby. This means that the computer could write two disks in parallel and when one disk fails the computer will be able to continue on the remaining one. If the computer failed, processing could be switched automatically by a "watchdog" switch to the alternate computer which could find out, from the disk, what was going on and continue.

2.3. Commercial Sector

The manufacturing industry consists of two interdependent sectors, commercial and conversion. Unfortunately, when investigating the efficiency or productivity of an industry/plant/FAS, the commercial sector is often overlooked or ignored in the analysis. Yet it is this sector, consisting of finance, sales, marketing, purchasing, and the general secretariate, that is the determining factor in the equation that says whether or not a given order is delivered to a client on the due date.

This last statement is seen to be true, when it is acknowledged that long before the conversion sector starts work on a given order, it (the order) has been the object of attention within the commercial sector. The order has to be acknowledged; any variable details sorted out, credit checks have to be run, materials and components ordered (and chased), and internal paperwork regarding schedules and delivery dates forwarded to the conversion sector by internal paperwork. This activity often occupies a lot of the timespan of an order within the FAS, and consequently limits the amount of time for the conversion sector to fulfill the order.

Analysis of the commercial sector indicates two major sources of inefficiency. The first is due to the often poor communication between the two primary sectors within the FAS, which leads to intersector conflicts regarding delivery dates and priorities. The second is the efficiency with which the commercial sector produces its output, namely, information. This output is generally in the form of paperwork and it is this latter inefficiency that is the major concern of this chapter.

It has been estimated that in the past century, productivity has increased in the "factories" by 1,000 percent, whereas in the "office" it has increased by only 70 percent.[3] For a given order, the external written communication generated by the commercial sector involves order acknowledgements, credit checks and approval, purchase orders, advice notes, delivery notes, invoices, etc. With the cost of having a "standard letter" typed, escalating almost daily it is not surprising that a major assault on the inefficiencies of the commercial sector is taking place, through the automation of office techniques.

The concept of the fully automated paperless office, with interoffice communications via electronic mail, automatic document retrieval, etc., is seen to be already viable through its precursor—the word processor. Unfortunately, in the general fear of automation, technology (in the case of word processors) has become confused with job function and therefore a precise definition of word processing is required. A conceptually-based definition that will stand the technological test of time is as follows:

Word processing is the production of syntactic text which may be retrieved and amended.

Therefore, it does not matter whether the text is produced by pencil and paper or whether it is produced on a million-dollar microprocessor-based system. Nowadays, however, the term usually refers to devices that have visual display screens (VDUs) of some sort, a keyboard, an electronic memory, and a printer.

The main purpose to which word processors are put is the preparation of "top" copies of documents. Through the keyboard, the text is entered simultaneously into the memory and onto the VDU, so that it can be observed by the operative. The advantage of these machines is that the document can be amended, corrected, rephrased, and changed in format in a matter of seconds. The word processors can also be personalized so as to automatically compensate for operative quirks. For instance if the user occasionally keys in "tye" or "thw" instead of "the", then the word processor can be used to search for these words and substitute the correct spelling when requested. Obviously this feature should be used with care since some character substitutions can form correct words, and replacement of all of those "words" in the text could result in a garbage letter.

The use of word processors means that a range of standard letters, paragraphs, addresses, etc., can be stored in memory, and when required, the appropriate part can be called from memory and the desired document then "assembled." The more sophisticated the word processor, the more it approaches the power and flexibility of "small" computers, yet even the simplest system can produce dramatic changes in the organization and economies of text preparation.

The cost benefits of a word processor system can be evaluated using different criteria. These can be an assessment of the reduction in the cost of producing the ubiquitous "standard letter" (usually by up to 80%), or where there is no standard letter or other measure by which office costs can be determined, then the benefits can be assessed by the reduction in "turnaround time," "error rate," or the increase in the quality of the type print.

Disregarding the method of rationalization, it is a fact that when word processors are introduced into an organization, large savings in proportion to the number of documents produced are achieved. As well as immediate cost reductions, word processors offer significant advances in producing clean top copies, rapid throughput of documents, and the updating of rapidly changing texts such as manuals and price lists. One widely acclaimed system in the U.K.,[4] with a staff level 50% of that prior to the adoption of word processors, achieved a productivity level 19% higher than in the previous arrangement. In other words the output per person had more than doubled.

Indications of the increase in productivity are a 12% reduction in the turnaround time for corrections and forms, through a 50% reduction in the turnaround time for letters and memos, to a 75% reduction for labels. The error rate, in terms of number of keys indexed, fell from 1:59 to 1:367. Operator morale also increased with turnover dropping from 35% to "zero" and attitude studies indicating that the operatives were much more satisfied than they were previously.

The fundamental purpose of the commercial sector is to process information. A large portion of the throughput time of a product is spent in nonproduction related activities—for instance, information processing, which while important does nothing to the value of the product. Consequently, increasing productivity within the commercial sector will increase productivity of the FAS.

Presently, office procedures tend to be human, and paper based—linked up with the files in one person's desk or the "personal understanding" that one person has with another. This means that the information holder can sometimes be holding the company for "ransom" because of reluctance to "share" the information.

Within the FAS of the immediate future, the commercial sector will be computer based with word processors, "electronic filing cabinets," electronic mail systems that will permit the transmission, receipt, storage and reading of textual messages that will (probably) never appear on paper. The local computer network will interconnect with a full range of printers, file stores, facsimile transmission devices, telex and other terminals, and all will be plug compatible. One of the major development areas will be in the field of FAX (facsimile transmission devices or facsimile exchanges), which because of its ability to transmit images such as drawings and signatures will become more of an universal electronics reader, merging with existing optical character recognition technology and evolving into a multipurpose digital camera.

The result of the introduction of sophisticated electronic equipment in the commercial sector is that communication between the FAS and its suppliers and clients can be conducted electronically, which removes the time occupied in dictating, typing, and mailing of letters. The information can be updated regularly so that the dynamism of the FAS is reflected by its communication system. Sales inquiries and marketing information can be analyzed by the appropriate computer programs instead of relying on a salespersons "guesstimate" of what may or may not happen. Accounts can be rendered instantly and where appropriate credit/cash transfer can be achieved by the computers themselves. Likewise, information regarding delivery and/or schedules will be available in real time, which means that nobody should get caught out. Finally, the microelectronic office can be linked to the other microelectronic systems so that the *total* FAS functions as an integrated whole.

Material Control

Flexible assembly systems are required to operate at the highest feasible productivity. Their primary function is to turn purchased raw material and components into value added goods. The output is dependent upon two associated facts. These are: (i) the materials and components must be present and correct for their conversion, and (ii) in order that the product's quality can be assured, these purchased items must be of the correct quality themselves. The prime cost of a product depends upon the amount of consumable materials used and the cost of all purchased materials used in the assembled product. Too much consumable material or too high a quality grade of material will escalate the true cost of that product, and reduce the profit level. In short, material control is an important aspect of the FAS.

Material control is a strictly managerial function and in its widest sense accounts for the majority of the throughput time span of a given batch of products. In order to improve the productivity it is imperative that these tasks (external to the conversion process) are performed at a high level of efficiency. Seven aspects of material control are discussed below.

3.1. Forecasting

Forecasting is the anticipation of forthcoming orders and/or manufacturing demand based upon historical information. While the FAS may have its order book filled for a few months ahead, the lead times for the procurement and delivery of certain raw materials and components used in the system's products may exceed the capacity of the order book and the delivery time for expected repeat orders. This means that the FAS has to estimate its future production program when placing orders for long lead items. In the past, forecasting techniques tended to be crude and required a lot of clerical effort to keep up to date. The advent of the microcomputer has given access to a wide selection of forecasting techniques (in real time) that can easily be used without the prerequisite of a Ph.D. in computer science. The techniques for which computer packages are available include: exponential, Box–Jenkins, correlation and lagged relationships, econometric models, and moving average and simple linear regression models.

In short then, the forecast indicates the usage of materials over a given timespan. In order that these materials can be used, they must be ordered and inspected prior to stocking. The delivered quantity is a function of the stock control technique used, the vendor's quantity/price breaks and associated delivery sizes, and finally the agreed delivery schedule. One aspect of material control that often gets overlooked is the rate of component and raw material delivery to the goods inwards inspection department (prior to entering the warehouse).

BRSL[2] determined that the delivery rate could be predicted thus:

> To make 300,000 of anything per year, the production per two-shift working day (240 days per year) would be 1,250. If these units require "n" of a particular component and those components are sold in minimum lots of "x", then the delivery rate per day will be . . . $1,250n/x$.

For a particular product consisting of 125 components of 71 types, the daily delivery equated to 156,250 items. This "average" delivery would occupy a volume of 35 cubic meters and weigh 17,500 kilograms. This calculation makes no allowance for scrap replacement or safety buffers against stockout.

3.2. Goods Inward Inspection

With items arriving daily at the above rate, the goods inward inspection will have to process the items prior to the next shipment. In other words, the maximum amount of time will be two shifts. Microelectronic-based goods inspection is fairly standardized nowadays with the advent of light pens, VDUs, and bar code or optical character recognition systems.

3.2.1. Bar Codes

A bar code is a pattern of parallel lines and spacings that can be read by a computer-linked optical scanning head. Simplistically, the pattern is read by passing a light pen (wand) across the code, moving the code past the reading head, or using a laser scanning device where the laser scans the code (with the coded label either stationary or mobile). The variation in contrast between the light and dark bars as they move relative to the sensor causes a binary code to be generated.

The bar code can exist in many configurations,[5-7] however Figures 6 and 7 show two tried and tested formats. The first is a set of variable thickness black bars where all of the adjacent "1"s in a character are blacked in. The second is a discrete code in which all of the digits are coded and the ratio of dark bar to light space identifies "1"s or "0"s. All codes have stop and start "guards" at either end of the pattern, which inform the computer the direction in which the code is being read. This means that the computer can either "reject" that reading or invert the data.

Many bar codes incorporate codes in "human readable form," so that they can be read by optical character recognition (OCR) systems as well as by a person. The

Figure 6. The Universal Product Code is the set of thick and thin bars printed on many supermarket items now sold in the U.S. It encodes 12 digits. Six of them, to the left of the "center guard pattern," are each represented by a light space, a dark bar, a second space, and a second bar. The other six, to the right of the center-guard pattern, are each represented by a bar, a space, a bar, and a space. The arrangement enables the computer at a grocery checkout counter to determine whether the code has been scanned backwards by the sensor at the counter. (In that case the computer inverts the data.) The 12 digits have various meanings. The first decoded digit, which also appears in "human-readable form" at the left of the pattern of bars, is called the number-system character. A zero signifies a standard supermarket item. The next five decoded digits (in this case 12345) identify the manufacturer of the item. The five digits after that (67890) identify the item itself. The last digit, which does not appear in human-readable form, serves to confirm that the other 11 encoded digits have been scanned and decoded correctly. It is the smallest number that yields a multiple of 10 when it is added to the sum of the second, fourth, sixth, eighth, and tenth decoded digits plus three times the sum of the first, third, fifth, seventh, ninth, and eleventh digits. Corner markings define the area that should be blank around the set of bars. (The set of bars was prepared by the Photographic Sciences Corp.)

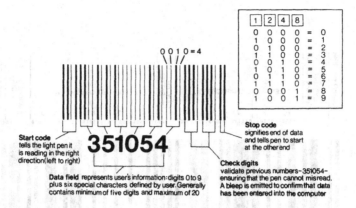

Figure 7. The Plessey unidirectional bar code—one of the many available codes—uses a binary coded decimal (BCD) format for the information in the data field, where each number is represented by four bits, or vertical bars. A wide bar followed by a narrow space represents 1, a narrow bar followed by a wide space represents 0. Using the BCD system, a 1 in any column represents the number shown in the box above that column and the total is simply achieved by adding. (Courtesy of *Engineering Computers*.)

OCR alpha numeric systems offer a high first-pass read rate compared with the regular bar codes. In addition, the use of a code that is easily read by the unaided human observer and a computer would seem to offer more benefits that a system designed solely for a computer-based recognition system.

Whichever code is used, they require special printing so that they are "compact," legible, and meet the regulations of size, position, and colors specified by the appropriate authority. This then assures that the "manufacturer" who uses the codes on products and the supplier/purchaser of the bar code reading equipment both know what to expect.

It is often the case that the code on the outside of a crate of goods will be different than the code on the individual items themselves. This is because the container has a lot more space for the code and consequently a larger, longer code can be used to offset the poorer quality printing and environmental damage that could result in poorer reading contrast.

The net result is that all items arriving at the FAS will be checked by reading a bar code. The information gathered as a result of this act will be used to generate all sorts of information relating to the "arrivee" and pass on information for management analysis. If the VDU fails/rejects, then the information regarding the delivery (using the OCR) can be entered by the goods inward inspector. The computer, when activated, will display details of the order, the level, and the type of quality control checks required before acceptance, as well as other pertinent information. Quality control is used to assure the conformity of the delivered goods to their specification. The specification can be their dimension, function, and/or visual appearance. For dimensional and functional checks the appropriate statistical quality control technique can be used, whereby a predetermined sample, based upon the importance of the component and the reliability of the vendor, will be used to determine the probability of defective items within the set of delivered goods. Where appropriate, 100 percent inspection, or a glance into the containers, will satisfy other criteria. With certain goods it is neither the dimensions nor the function that is important, but rather the appearance of the items. Here the acceptance checks are all visual, thereby subjective, and are all fully 100 percent carried out to check for flaws or damage to the external surfaces of the components.

When the goods have been inspected, they are either accepted or rejected. In the latter case they are returned to the vendor with documentation stating the reasons for rejection. When goods are accepted they have to be transferred to the warehouse, which because of the variety of shipping containers means that they have to be transferred to containers that are compatible with the handling and storage equipment of the warehouse. While many containers and types of warehouse storage equipment have the same "standards," the sizes that make economic sense for intercompany shipment are not often logical for use within the company itself.

The computer will indicate to the goods inwards personnel the type and size of the container to which the goods are to be transferred. The system will then supply the appropriate containers, and the delivery container will be broken open and the contents transferred. Each "in house" container will have its own unique identification code—this could be a bar code or a binary code plate—that is rigidly affixed or indelibly marked on the container. Assuming the use of bar codes, each container will be placed onto automatic weighing devices that will determine the number of components within that container. This means that the use of the computerized recognition system simultaneously identifies a particular container as holding a specific number of a given component and it will allocate that container a unique storage space within the warehouse. This ensures that, when required, the computer can identify the storage space(s) of a given number of a particular item.

While the delivery rate of goods can be averaged or predicted for a particular time period, the availability of a particular item at a specific time depends upon the stock control policy implemented.

3.3. Stock Control

Stock control occupies a Jekyll and Hyde position within a company since too much stock can seriously effect the company's cash flow and too little stock can prevent products from being processed through stockouts of (classically inexpensive yet critical) components. Purchased materials have either dependent or independent demand profiles. Independent demand items are those that cannot be inferred as belonging to any specific future product. They tend to be common fastening devices or replenishment of regular raw materials that are used for a variety of applications. The method of stock control for these items has four forms.

3.3.1. Independent Demand Item Control Techniques

The four stock control techniques are all different and are used to satisfy different criteria and operational conditions. The first is a very crude yet effective technique for low-value–large-usage or medial-value–low-usage components. An economic model follows that aims to minimize the incurred costs of stock control, but ignores entirely the risk and cost penalty of stockout. The last two models look at stock control as a dynamic system and offer two alternative methods of minimizing the *total* risks, and consequently costs, based upon a specified risk of stockout.

3.3.1.1. *Two Bin.* This technique is used for inexpensive, unimportant items. Two bins of the item are stocked, one of which is used to supply requirements and the second is used as a buffer against stockout. When the first bin is empty, the second is used to satisfy demand while a replacement bin is ordered and delivered.

3.3.1.2. *Economic Order Quantity (EOQ).* This technique is based upon minimizing the total cost of stocking an item. The optimal order quantity is calculated using the annual demand for the item, the cost of placing the order(s), the desired safety stock, and the cost of storage per annum for the same item. Figure 8 shows the relationships of these different parameters, and the formula for determining the EOQ is given below:

$$Q = \left(\frac{2C_0 A}{C_t} \right)^{1/2}$$

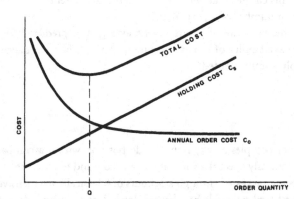

Figure 8. Economic order quantity (Q) represents the minimization of the total incurred cost based upon the holding cost, the cost of ordering, and the annual demand. It takes no account of the call-off rate or many of the other factors that identify the ideal of the FAS.

where C_0 is the cost of placing an order, C_s is the cost of storage, A is the annual demand for an item, and Q is the optimal order quantity.

This technique takes no account of the call-off rate for the items, neither does it include any factors that take account of variable delivery time by the vendor or the desired risk of stockout that is specified by user of the method. In short, it is a deterministic technique that has limited use in the dynamic environment of the FAS. The following two techniques are stochastic, whereby probability values are given to both the usage rate and the vendors delivery period.

3.3.1.3. Reorder Level Policy. This technique computes a fixed quantity that, if ordered when the existing stocks of the item reach a specified number, will satisfy the demand requirements and ensure that the specified assurance against stockout is honored. Figure 9 shows this technique graphically, and the formula for determining the order quantity is given below:

$$M = DL + \eta_\alpha \sigma(L)^{1/2}$$

where M is the reorder level, D is the demand over time period, L is the lead time, η_α is the standard normal for a specified stockout, and σ is the standard deviation of demand.

Where the lead time varies, the method is to calculate M for each of the lead times and then, by means of an iterative process, test for the value of M that gives the probable stockout value that is desired.

3.3.1.4. Reorder Cycle Policy. This technique computes the order quantity that is required to replenish the current stock level to the desired (optimal) stock level, at a fixed time interval. Like the previous technique, this time-based stock control system will satisfy the demand requirements without exceeding the expectations of no stock-

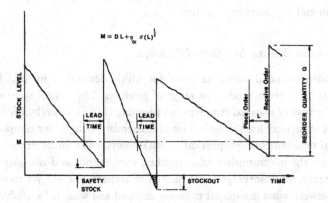

Figure 9. This stock control technique is based about a given and known quantity, Q, being ordered when the stock level reaches M. Variation of the lead time of delivery can cause stock-out, but this is limited to a known probability.

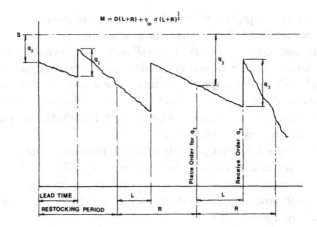

Figure 10. This stock control technique is based around a given and known period of time, at the end of which the difference between the desired and actual stock levels is computed. An order for this quantity q_i, is then placed with the lead time resulting in the stock level being increased or stock-out occurring if the lead time is extensive and the demand for that time period is excessive.

out. Figure 10 shows this technique graphically, and the formula for determining the order quantity is given below:

$$S = D(L + R) + \eta_\alpha \sigma (L + R)^{1/2}$$

where S is the preferred stock level and R is the restocking periods. All other symbols are as in the previous technique, and variable lead times are dealt with in a similar manner. For indepth information of these and other independent demand item control techniques, the reader is advised to consult any of the many specialist publications on stock control and inventory management.

3.3.2. Dependent Demand Item Control Technique

Dependent demand items are those for which demand is inferred by future requirements for a given product or range of products. The control system used for these items is that of material requirements planning (MRP), whereby the forecast or requirement at product level is broken into subassembly, bought out component, and raw material requirements. Simplistically, MRP analyzes the future demand requirement, deducts the uncommitted subassemblies, components, and raw materials that are in inventory, work in progress, or have been ordered but not yet received. The difference between what is required to satisfy demand and what is "available" is then ordered from the various vendors. If the vendors' delivery dates adversely effect the delivery schedule of the products, then the MRP controller can advise management.

Likewise, the scheduling of the conversion sector and its capability can be checked, revised, and any conflicts resolved. This stock control system is ideal for a FAS, since it operates in real time, is very dynamic, and is of course, computer operated.

3.4. Pareto Analysis

Any item has two costs to the FAS. The first cost is the purchase price and the second is its "uniqueness cost." The classic example linking these two costs together is where the stockout of a ten-cent washer prevents the delivery of a multi-thousand-dollar machine. The cost of the washer is then equal to the value of the invoices that cannot be sent, plus the penalty clauses that may be invoked by the client. Pareto analysis is a method for classifying items in terms of their contribution or responsibility for a given criterion.

If a group of items are analyzed against a series of criteria, it will be observed that a small proportion of the group are responsible for a high percentage of the given criterion. For example, if a group of products are compared against turnover, it will be found that approximately 20% of the group are responsible for 80% of the turnover. Using Pareto analysis classification, the 20% will be allocated as "A" category products, with the remaining group members being allocated to "B" or "C" categories dependent upon their influence on the criterion. It is important to remember that the analysis should be applied to each group several times, using different criteria, before allocation of the ultimate classification, since—dependent upon the criteria used—a particular group member could be classified "A," "B," or "C." Hence, final classification is often a managerial decision. It therefore follows that if stringent stock controls were applied to the "A" category items, lesser controls to the "B" items and few if any to the "C" category, then control over the whole group would be achieved with emphasis where it counts.

The 20/80 split seems to be fairly consistent no matter what activities or items are compared against what criteria. It therefore makes sense to use Pareto analysis whenever appropriate for testing the importance or contribution of any activity or item within the FAS (see Figure 11).

3.5. Consumable Materials

Many assembled related tasks use consumable materials, such as welding rods, hot melt adhesives, sealants, staples, foams, etc., without monitoring the usage rate in terms of that applied or wasted. Today's world of high-interest money, energy conservation, and pollution control means that consumable materials should be applied in the correct amount in the correct place to achieve the desired function.

Many products use paper or cork gaskets in order to achieve a gas tight joint

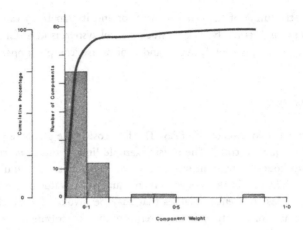

Figure 11. This simple example of Pareto analysis shows that about 95% of the components examined during the BRSL study weighed a maximum of 0.2 kg each.

when clamped in position between two flanges. The manufacture of these solid gaskets is expensive, they are easily damaged, and because of economies of scale they are made in large quantities with the corollary that a lot of storage area is required. An alternative to these solid gaskets is the use of a sealant dispensing machine, which can be integrated into an automatic assembly line. These dispensing units can lay simple or complicated bead paths, and are shape selectable through the use of sensors or switches. The units are fitted with 25-kg drums of sealant or adhesive and their percentage waste at changeover is minute compared with that wasted in cartridge systems. The cost savings for these dispensing systems is based upon the cost of sealant versus the solid gasket, the cost saving of not having to stock gaskets against stockout, and the cost benefit of being able to change the gasket shape almost at will.

Welding rods present another case in point. Manual welders do not always achieve the correct penetration or weld path. They also cause a lot of splatter and throw away partially consumed welding rods. The use of automatic welding equipment, and especially robot welders, ensures that the penetration is obtained and that it is consistent for the weld length and the batch of units being welded. Also, because these welders use rolls of wire instead of individual welding rods, the percentage of waste is small compared with that of the manual welder. Likewise, the cost per unit length of weld in terms of material alone is less with the nonmanual system by virtue of the cost savings—through the economy of scale—of buying reels and not individual welding rods.

Many products are designed without thought to the use of consumable materials within their structure. A prime example is the use of fasteners where a hole is drilled and is then filled with a screw or pin. If the local assembly task could have been done by say, ultrasonic welding, snap-in integral "fingers" or adhesives, then a cost saving relating to the fastener and its storage charges would be made. It is also possible that the assembly task itself would be easier, and so contribute to a lower prime cost.

3.6. Inventory Management

This is a feedback control system that allows for assembly against sales operations, rather than assembly against forecast operations to be realized. When products are completed, they are (usually) crated and entered into inventory to await call-off for delivery to a client. Analysis of the inventory function will give the storage time for particular products. It follows that products that do not move should be reexamined for their viability as a marketable product. The inventory turnover should also indicate which products should be assembled, namely, those that will only replenish and not build stock.

One measure of an FAS's efficiency is that of inventory turnover. The higher the rate of inventory turnover, then theoretically the shorter the time that any particular item spends sitting in the stores absorbing storage charges and contributing nothing to the FAS's cash flow situation. The high level of interest rates and the cost of storage can mean that the cost of stocking materials, components,and products can be around 20%–25% of the inventory's value per annum. It has been estimated that for every 1 percent reduction in inventory value, the return on investment would be increased between 1 percent and 1.5 percent for a typical manufacturing company.[8]

In order to achieve savings the FAS must know what it is trying to control. The key to control of stock levels and usage of materials are the techniques and disciplines of material requirements planning (MRP) and production control systems. The next requirement is to have control of the items themselves such that they do not get "lost" or overlooked in the storage areas of the FAS. The removal of randomness from stock control is achieved through the use of computerized systems that track the materials and components from the time that they enter the FAS to the time that they are incorporated into products, which are themselves then entered into the computer records. In order that control is maintained over physical items it is necessary that they be entered into the storage area by a system that is entirely under the control of the computers and in which people take little or no part, for once an uncontrolled element enters the system, errors can and will occur.

3.7. Computer Controlled Warehouse

The choice of a computer controlled warehouse is a very subjective matter and it is of benefit to follow BRSL's[2] logic in deciding and specifying the warehouse that matched their needs.

The stock control philosophy as specified (in the brief) was to hold at least eight weeks supply of each component. While computer controlled warehouses are generally associated with increased stock control, and consequently a reduced need for buffer stocks, it was considered prudent (in the U.K) to hold this amount of stock to counter problems with vendors. At a later stage, when the "products" were defined and the assembly system stabilized, then the "vendor" problem could be examined and resolved. The stock control philosophy could then be amended to one that was more in tune with an FAS's image.

A number of other assumptions were made so that the movements and storage volumes could be used as a basis for estimating the size of the storage volume required and the rate of retrieval. These assumptions were:

   (i) The existing typical component should be used to simulate the number of components handled;

  (ii) The configuration and work schedule and consequently demand schedule on the computer controlled warehouse should be based on this same product;

 (iii) There should be only one bin size. The analysis would allow for varying the actual bin chosen, but not variation of bin size within an analysis;

 (iv) Within a bin, goods could be stored three ways:

    (a) Loose for very big parts;

    (b) Loose in sub bins, ⅛th the total bin volume;

    (c) On molded trays;

  (v) Each day should be divided into two eight-hour shifts and all automatic devices should continue to work through lunch and coffee breaks; and

 (vi) Only one type of component or subassembly should be stored within one bin.

One recurring subject was the need to specify and standardize the size of the basic material handling bin. The following factors influenced the decision making:

   (i) For ergonomic reasons, it was desirable that the bins be of a size that permit at least four (bins) to be placed around an operator at a work station such that they were all easily accessible;

  (ii) For material control reasons, it was desirable that small amounts of components be issued without losing the integrity of those still held in the bin, while at the same time ensuring that the bins were sufficiently utilized;

 (iii) Since "large" parts constitute 70% of the movements, as large a bin as is possible was desirable;

 (iv) It was also desirable that the bins be of a size that would permit them to be easily picked up by people. This meant that their loaded weight could not exceed 10 kg, with the result that bins containing small dense parts were underutilized.

From these factors a bin size of 0.4 m wide × 0.5 m deep × 0.3 m high was chosen after a number of computer simulations and in-depth calculations.

Using this bin size, the various components were allocated bin or sub-bin status, and the number of bins and bin movements needed to satisfy production demand was determined. The analysis took cognizance of the fact that some subassemblies would be passed between work stations in a given work area, while others would be returned to the warehouse from the "last" work station and retrieved from the stores at a later time.

The calculations indicated that under ideal circumstances there would be 16,355 bins required and that there would be 894 deliveries per day or approximately one per minute. The most obvious wrong assumption—that all bins would be full—was deliberate. To make any other assumption would be to force a fairly subjective error into the analysis. The operational policy was to keep full bins in the store and to use partially full bins for shipments. Where possible, and excluding sub-bins, whole bins are to be used at the work stations. When new goods arrive at goods inward they are to be stored as full bins. Additionally, no allowance was made for errors or scrap, neither were any allowances made for goods inward or goods outward requirements.

The bin deliveries will not be evenly spread around the various work areas. The final assembly areas account for 35.7% of movements and a dedicated PCB line accounts for 39.2%. Unfortunately, this means that the elevators and cranes that service those two areas will be subject to far more pressure than those servicing other work areas. It should be noted that there is a "domino" effect from certain bin movements, in that while the finished goods directly account for 11.7% of movements, they also generate a number of crane and elevator movements. These "extra" movements are the result of the difference between the number of bins needed to supply a work station with parts, and the number of bins required to remove the assembled units from the same work station. Therefore, if the work stations are to be kept supplied with material, then the number of delivery bins will always be significantly larger than the number of collection bins. The need for special trays and/or subdivisions for certain components and subassemblies would mean that those "dedicated" bins would not be reusable on a general basis, albeit that they could be manually fitted and removed externally to the computer boundary of the warehouse. Finally, it was determined that of the grand total of 16,355 bins required to satisfy the storage needs, only 894 (5.5%) would move on any one day.

From the beginning, BRSL assumed that an automated solution would be preferable to a traditional manual warehouse and material movements system. This assumption was not based on an "automation for its own sake" approach, but rather on the idea that the traditional methods would increase the occupied area and running costs of the total FAS, without providing an efficient support for the unit. Also, "transactional accuracy"[9] is increasingly being identified as the primary requirement for stock control. The phrase "garbage in, garbage out" has long been associated with computers. But nowhere are the dangers of informational garbage more apparent than in the warehouse, since if a component is put into the wrong storage bin and/or that bin is randomly placed in the warehouse, there is no way of finding that component and/or correcting the error short of taking a manual physical inventory of each and every bin.

Given the pressures toward improved warehouse service at ever lower cost, the control systems of the warehouses will have to be more dynamic than they have been in the past and function with lower stock levels. They cannot do this without more accurate information, and there is now a greater awareness of the potential for inaccuracy that exists within a conventional warehouse, as well of the limited level of accuracy that can be expected of the human operator.

Since the warehouse is associated with almost every other system in the FAS, there were a number of interactive questions that had to be resolved. One industrial practice is to free issue work stations with small cheap items in bulk. It was determined that this system should be continued after modification to suit the computer system, since it is a logical procedure to bulk issue this category of parts rather than to issue them on demand. The transfer to a computer-based dispersal system permits inventory control to be maintained. It is also acknowledged that batch size would be a function of the container size as well as the usage rate.

Part integrity was another concern in that for reasons of quality assurance, it was desirable to maintain the integrity of a batch of components. In this sense a batch is taken to mean components that were delivered in a single consignment and are identifiable by both manufacturer and date. This means that "new" components should be stored in fresh bins, and that the bins should be reidentified by passing them through goods inward (to establish the correct quantity) and the system should be able to issue partially full bins. It must also be possible to identify and amalgamate partial loads at the goods inward station. An associated concern was that it should not be possible to remove goods from the stores without either the store control system being aware of it or without authorization. Concern was further expressed about the number of parts that should be issued from the stores. In essence it was a case of whether full containers should be taken to the work areas and the appropriate quantity removed prior to the remainder being returned to the warehouse, or whether only the required amount should be presented to the work area. The concept of kitting was discussed, but was not unanimously accepted as being the correct solution.

Dispersal was the final concern that had to be resolved. The requirements were that the goods should arrive at the work stations as needed. Ideally, this would mean a system that could identify the particular work station's requirements in advance and move any part anywhere as required. While this was the goal, it was realized that the degree of flexibility implied would probably result in a system that was too complicated and sophisticated, and consequently too expensive, to justify. The net result was a compromise, where the designated system would work excellently under predictable conditions, fairly well in emergencies, and not at all if pushed to its limits—as, for example, if all work stations attempted to change over simultaneously from one product to another.

Based upon the simulations, it was obvious that whatever dispersal method was used, a rapid rate of retrieval would be necessary. While the analysis only considered movements of goods from the warehouse to the various work areas and back again, there are three other motions that have to be considered in the total solution:

1. Movement of goods from goods inward to storage;
2. Movement of finished goods (subassemblies or products) to storage, and from storage to dispatch; and
3. Incidental movements such as re-entry of partially filled bins, and shuffling of bins to optimize locations within the warehouse, and so on.

The determined rate of transfer of bins, namely, 1/min, relates to a unidirectional motion, with goods arriving and departing from the stores or a work area being classified as separate movements.

3.7.1. Automatic Storage and Retrieval Systems

The system generally used for computer controlled high-density storage systems is the automatic storage and retrieval system, or AS/RS. This system consists of high-speed stacking cranes that can travel horizontally at up to 250 m/min and vertically at 40 m/min. They are usually confined to storage aisles, although systems are under development which will allow the cranes to operate inside and outside the storage area. Currently, cranes operate in gangways that are only slightly wider than the "loads" and they are usually mounted on floor rails. While the majority of the systems are below 20 m in height, there has been a shift to very high structures exceeding 35 m in height. One particular advantage of high bay storage is that the storage racks can be used as the structural framework of the building. This type of clad rack structure (for certain countries) brings corporate tax benefits, since for tax purposes, the integrated system is regarded as a machine and its whole cost can be written off in the first year. In the case of a conventional system, as many as 13 years could elapse before the full allowance is received.

The system that emerged as the most suitable was that of Vickers/Supreme's "Conserve-a-trieve,"[2] which basically consists of facing banks of storage containers, up to 9 m high, between which run electronically controlled mini-load stacker cranes. Each computer controlled crane travels to the appropriate position in the aisle. At the same time (if necessary) as the crane is moving horizontally, a platform on the crane moves vertically. When the crane is at the correct position, a container is pulled onto the platform by an extraction device. The container can then be moved to another location or taken to an end of an aisle automatic interface to a conveyor or picking station. To give the system greater speed and flexibility it was also decided to situate high-speed elevators at the tops of certain locations along the store racking so that the containers on these locations could be moved directly to the work area situated directly overhead.

The reasons for choosing the stacker crane style of stores were the speed of access plus the density of storage and the flexibility that would accommodate product changes. The Conserv-a-trieve system was attractive since it gave total automated control with the possibility of adaptation for individual needs. It had also been marketed for several years and had a proven applications record. Finally, the system was modular in design and could easily be adapted or enhanced. Figures 12 and 13 and Table 1 compare various storage systems by a number of criteria.[10]

Taking into account errors, unforeseen requirements, and possible future expansion, BRSL determined that the store capacity should be three times the calculated number of bins. Rounded off this gave 50,000 bins, and the physical size of the store for this number of units was determined as follows:

Height of racking: 9.500 m
Height of bin: 0.300 m
Height of "slot" (0.061 m, clearance and racking): 0.361 m

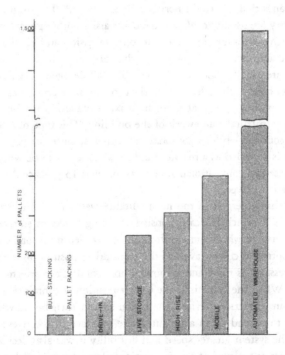

Figure 12. One of the comparative measures of storage systems is that of pallet capacity. This figure shows that the automated warehouse has the highest density of storage with an almost four-fold edge over the next most efficient system. (Courtesy of *Engineering.*)

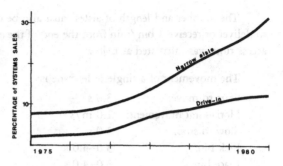

Figure 13. The movement toward high-density storage systems is shown by this graph in which it can be seen that the sales of narrow aisle systems (as a percentage of total sales) has changed dramatically over a five-year period. (Courtesy of *Engineering.*)

Allowance for structure:

Distance to bottom shelf: 0.584 m

Distance from top shelf to "ceiling": 0.249 m

Tie bars loss of 9 percent approx.: 0.864 m

Splicing posts, loss of 6 percent approx.: 0.533 m

Total losses due to structure: 2.230 m

Total available height for storage: 9.500 − 2.230 = 7.270 m

Therefore, number of bins (in height): 7.27/0.361 = 20

Number of bins per bay, assuming that bins are stored two deep on either side of an aisle: 40

Therefore, number of bays to store 50,000 bins: 1,250

On the assumption of eight aisles, the number of bays per aisle would be: 1,250/8 = 156

Assuming that five bays per aisle would be used for maintenance and that the width of the bin slot is 0.461 m, then the length of each aisle is: (83 × 0.461) + 0.400 = 40 m (approx.)

Table 1
Comparison of Various Storage Methods by Criteria Over Elementary Storage Techniques

Criteria	Bulk stacking	Pallet racking	Narrow aisle high rise	Drive-in racking	Storage platforms	Reach truck racking	Pantograph racking	Mobile racking	Live storage
Cube utilization	Fair	Good	Excellent	Excellent	Very good	Good	Very good	Excellent	Very good
Accessibility of unit load	Restricted	Excellent	Excellent	Poor	Good	Excellent	Fair	Good	Poor
Order picking	Restricted	Excellent	Good	Poor	Good	Excellent	Fair	Very good	Excellent
Speed of throughput	Fair	Good	Good	Poor	Fair	Good	Fair	Good	Good
Load crushing	Probable	Nil	Nil	Nil	Nil	Nil	Nil	Nil	Possible
stability	Poor	Good	Good	Good	Good	Good	Good	Good	Fair
Stock rotation	Poor	Very good	Very good	Poor	Fair	Very good	Fair	Good	Excellent
Potential savings	—	—	180%	80%–90%	—	20%	50%–60%	90%	90%

Reprinted from *Engineering*, April 1982.

The number and length of aisles must also be compatible with the requirement to deliver or receive 1 bin/min from the end of the aisle conveyors and elevators. The access time was calculated as below:

The movement of a single aisle crane is:

Start to move:	3.0 s
Horizontal movement:	1.0 m/s
Slow to stop:	3.0 s
Pick bin:	3.0–4.0 s
Place bin:	3.0–4.0 s

The time to move half the length of an aisle is:

Start to move:	3.0 s
Move 20 m:	20.0 s
Slow to stop:	3.0 s
Pick bin:	4.0 s
Start to move:	3.0 s
Move 20 m:	20.0 s
Slow to stop:	3.0 s
Place bin:	4.0 s
Total	60.0 s

Assuming one movement per minute per aisle, the stores processing rate covers that needed for the FAS. These average movements of the high-speed cranes can be further reduced with the output point at mid aisle and with the delivery and retrieval areas overlapped. The most critical time for the operation of the storage system is when product changeover occurs in the work areas. Given that the limiting factors are the high-speed elevator and the two delivery points per aisle, it is difficult to see that each point could be serviced more than once a minute, which means a maximum of two deliveries per work area. The analysis indicates that a maximum of 120 deliveries and removals (though overlapping) per hour would be adequate and that phased work area by work area changeover would not be disruptive. It is also anticipated that some interstore operations, such as transferring parts closer to their final work areas, would take place in the unattended third shift of the day.

In order to increase storage density and help reduce maintenance needs, it was decided to store two bins at each location on a single 1.00 m × 0.40 m tray. These trays would travel no further than the top of the high-speed elevators from where the bins would be removed or added. The bins would be of a less expensive design than if an alternative transit method were used, since they do not have to have a mechanical key into the rack. From a philosophical point of view, the warehouse unit consists only of the tray, and all crane movements would be based around this "unit." However the "unit" for control and inventory purposes would be the bin. Moreover, all trays would contain bins—not loose parts—and each bin would contain only one type of component, although the two bins on a tray could contain different components.

The two conveyors at either end of the aisles would be located at the lower level of the racking. This together with the fact that input and output are at different ends of the FAS should facilitate deliveries to and from the FAS. The conveyors also include spurs to enable trays to be brought to and from specified working bays. Eight bays would be available at either end of the store. At the input end they will be used for either breaking and packing or for quality assurance and inspection. A VDU will be available to ensure access to the computer system by the personnel for all functions conducted in this area.

Further, storage will be available in the form of palletized storage and racking, which is serviced by fork lift trucks. This storage can be used for large tools and any parts that are larger than those currently foreseen. It will also be useful as a temporary storage area for unprocessed incoming goods which require inspection before loading into the storage system. It could also be used for storing packaged finished goods that are arranged as orders for clients. Two conventional elevators between the floor containing the work areas and that containing the storage and input/output areas completes the computer controlled warehouse configuration.

The material distribution is achieved by moving the trays from the store, onto the high-speed elevators, and up to the end of conveyor stations belonging to the work area directly above. Each work area is serviced by a two-level conveyor system that encloses three sides of a rectangle that defines the perimeter of the work area. At each end of the conveyor will be a double lift, servicing both layers of the conveyor. While one side of the "double elevator" rises the other side falls—so inputs will always be taken to one level while outputs will be taken from the other. The trays will come up and have their bin(s) removed, they will then progress to the output station where any bin(s) can be added. The tray will then be lowered automatically by the double elevator.

Bins will progress around the conveyor until they reach an appropriate exit point from where they can be automatically extracted and placed into the stores, where prior to being entered into a specific "pigeon hole," they will be identified by their bar code, weighed, and the information entered into the computer records. One problem that arose in the past was that of allowing work stations access to more material than they required. The minimum that can be removed is one single sub-bin, since it would be difficult to automatically control inventory if less were allowed. One method of maintaining control is to automatically weigh bins and so determine how many components have been removed. Unusually large removals, especially of expensive parts, could be identified immediately and management alerted to the (potential) problem. Alternatively the sub-bins could be removed by a pick-and-place device outside the actual confines of the manual work station.

3.7.2. Computer Control Aspects of Material Control

3.7.2.1. Warehouse.
The Conserv-a-trieve system, and indeed any comparative system, is supplied with a certain amount of computer software and hardware. The

contents of the store should not be visible and an automatic record of where all bins and trays were placed should be kept. Prior to describing the system in functional terms, certain concepts must be cleared. These are that each bin will have a unique code, as will each tray. Each location will be described in the terms of its coordinates (e.g., aisle 3, left-hand side, third bay down, fourth location from the top). The identity code on the trays and bins will be permanently attached, indelible, and self-destructing if any attempt is made to remove it. The numbers will be both human and computer readable and be capable of being read by either a fixed sensor or a hand-held wand.

The command to "move" something in or out of the warehouse will be generated at a higher level in the warehouse computer. A work area will identify the need for a particular component and tell the computer, which will then ask the warehouse to move it to the work area. It is possible for the work area to also directly instruct the warehouse, but normally the request comes to the warehouse computer to move a certain amount of components to a particular work area. The computer then has to determine where the components are stored and which location is most suitable for the particular work area. This presupposes that the database held in the warehouse computer holds information on all components and locations, as well as all locations of each and every tray and bin. While this assumption is correct, it helps explain the complexity of the operation.

The next decision is how to route the goods to the work area. This involves parameters relating to what cranes are in use, the urgency of the request, which bins contain the "oldest" suitable components, and so on. Fortunately, the decision rules for this operation exist and a simulation derived and proven algorithm satisfies the requirement.

When the warehouse computer has decided upon the location "from" and the location "to," including the two input and output conveyors, it will issue the command to move the goods, plus their priority. While this decision making will be performed by specially written software, everything afterward will be done by, and be the responsibility of, the software supplied by the manufacturer of the warehouse. For this reason, it was necessary to maintain the integrity of the supplied software, and for implementation reasons, it was recommended that there be two computers. The first is the "customized unit" for controlling the warehouse, and the second is a separate, standard unit that is provided to operate the warehouse.

The command to move the goods is translated by the "second" computer into instructions to move the crane to a specific coordinate, pick up the goods, and then move them to the elevator system. There will be an interlock system that prevents hardware contention with the work area elevator system. The main warehouse computer will "know" where the bins are, or it will know where they were last dispatched and what they contained. The computer will learn of the true contents of the bins when they are returned for reentry.

3.7.2.2. Goods Dispersal. Each work area will have its own computer, which will oversee the local dispersal around the conveyor. The conveyor itself, the elevators, and the interlock between the conveyor and the work station storage system will be microprocessor controlled. The work areas however will control their actions functionally and coordinate their activities. When a tray is placed on an elevator, the warehouse computer will inform the work area computer that it is there, what it contains, and in what bin number the components are held. The work area computer then instructs the elevator to rise, and it informs the conveyor system what bins to remove and where it is to be routed. At the down elevator will be the automatic checking station for checking bin and tray identity (and weight if required).

3.7.2.3. Goods Inward. The personnel at goods inward will communicate mainly with the FAS computer, from where they will check deliveries, and be advised on the packing styles and test levels, etc. The warehouse computer will only be involved when the goods are ready to be placed in the stores and/or more bins are required. Each packing station and quality control station will have access to a VDU, as well as on-line weighing and checking devices.

3.7.2.4. Goods Outward. This will be similar to the goods inward in that beyond the actual physical movements of goods, most of the computer control will be from the main FAS computer. Dispatch and invoicing documentation will be printed automatically.

3.7.2.5. Security. The warehouse will be inoperable if the software or hardware fails. The warehouse computer will once a day print out a "map" of the stores and the location of the components. It will also cross check its files with those of the main FAS computer which keeps the master inventory. This cross check will be in volumes and/or quantities, and not in locations, which will be unique to the warehouse computer.

3.7.3. Implementation of the Computer Controlled Warehouse

The total materials distribution system represents a very large amount of complex automation and computer control. As such it would be unrealistic to try and implement it as a single entity. The warehouse itself—racking, etc.—together with the manufacturer's software, forms a single proven entity. All of this could be installed in the first instance along with the high-speed elevators, together with the input and output conveyors. However, the software would only support requests to move material between locations, so the contents map would have to be maintained external to the stores.

The work areas form a separate system that can be developed asynchronously. However, it is possible that they could go live without an automatic conveyor system. The elevators would then have to be operational, though not necessarily microcomputer controlled.

The goods inward and goods outward can work through a terminal through which the personnel could enter brief point-to-point commands. All other controls can be added as they are developed, with the original manual interfaces left in place to give security in case of system failure.

3.7.4. Costs of a Computer Controlled Warehouse

The following costs do not include those of the higher level computer and software control, since these are added into the total costs for the computer system given in Table 23 in Section 13.3. The costs (given below) were derived from the cost of each stacker crane, the cost of associated bins and racking, the cost of the conveyor system and fast elevators, and the cost of installation and commissioning:

1. Each stacker crane costs $120,000. Therefore, eight cranes will cost $960,000.

2. Each aisle is 40 m long and contains 83 bays on each side. Each aisle is serviced by one crane. The cost of a "twin bay" for both racking and bins is $3,000. Therefore, for eight aisles, the total cost will be $1,992,000.

3. Two conveyors, one for each end cost $40,000.

4. Fast elevators, one pair per work area cost $480,000.

5. Pallet stores and fork lift trucks cost $240,000.

6. Specially molded trays and containers cost $10,000.

7. Installation and commissioning at 20% of aisle and crane costs cost $590,400.

The total cost would be $4,312,400.

Work Centers

The heart of the FAS is the conversion sector, or work center, where the raw materials and purchased components are transformed into value added goods (or scrap). In order that this task is accomplished in an efficient manner, it is necessary for the time span of action within this sector to be as short as possible. The FAS is designed as a uni-directional flow path of activities and consequently the conversion and assembly tasks are achieved at a highly efficient rate. Productivity is also encouraged by the flexibility of the basic element of the work center—the work station. These "assembly modules" can be modified, reconfigured, and rearranged into "active" work areas containing only those work stations that are necessary for the tasks to be performed.

Traditionally, assembly related tasks have been a manual process and productiv-ity has been a function of the time spent by a particular operator performing a partic-ular task. When any function is human paced (i.e., semiautomatic machines and pro-duction lines, where if the human operator is not present, the process is not performed or completed), the productivity index can be as low as 40%.[11] As assembly processes become more and more automated, their efficiency rises to the 85%–97% range, depending upon the degree of human involvement or dependence in the process.

The human operator as a productive element within a FAS is not at all satisfac-tory, as they are inconsistent, unreliable, and expensive. The desired attributes of an "assembly medium" are those of judgement, dexterity, strength, and flexibility, all of which are lacking or diminished in people, as their judgement is fallible, their dex-terity is limited, their strength quickly fades, and their flexibility is not uniform.

The economic realities of the marketplace have necessitated that the conversion sector supply products of consistent quality, reliable functional operation, and a rea-sonable cost; further, that the output of these products be of a predictable nature so that consumer demand can be satisfied. The conflicting parameters of quality, quan-tity, and cost have promoted the development and operation of automated systems within the manufacturing industry. While automated machines and processes have existed for years—take, for example, the case of the tobacco industry, where for some time cigarettes have been produced on high-speed dedicated assembly machines—these (primarily) hard automation solutions are no longer compatible with the present-day socio-economic environment.

In today's world, many products exhibit a short life cycle, which means that the manufacturer's production system must be dynamic so that it can react quickly to changes in market demand. The need to balance product output against the fluctuations of market forces has promoted the development of flexible assembly systems. While these systems are primarily automated, they can enclose manual or semiautomated features within their confines. The flexibility of these systems is limited only by their parameters of geometry, power, and function, though within these constraints they can perform an almost infinite variety of tasks.

The choice between the three assembly methods of manual, flexible, or hard (dedicated) automation is made (objectively) on the basis of price per assembly produced and/or the cost and "rigidity" that is specified for the work area. Manual operations are used when either the task is too complex or too fragile for a machine, or when it is determined that a human operator makes economic sense. Flexible automation is used where an automatic system is desired for a product that could and will change in shape and size. As such, the system will have large power capacity and functional parameters, whereby it can accommodate many products (that may not yet exist) within its operational envelope. The third method is that of hard or dedicated automation, which is a fixed rigid system or machine for processing one product or performing one task (usually in vast numbers). An objective method of choosing between one assembly method and another is that of econometric graphs, which compare the cost per unit for various production quantities. Figure 14 shows one such graph. As can be seen, the cost of the manual process does not vary at all (piece work), whereas that of the flexible assembly is lowest until the quantity exceeds that of one machine. At this stage the hard automation system, which usually has a higher production rate than that of flexible automation, becomes the most cost-effective solution.

In order that the work station can perform the desired assembly tasks it must be serviced by (usually) hard automation devices so that it is provided with components

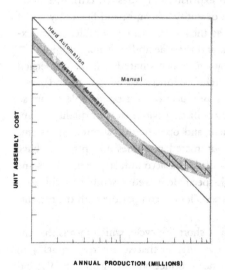

UNIT ASSEMBLY COST

ANNUAL PRODUCTION (MILLIONS)

Figure 14. This econometric graph shows the comparable cost per unit assembled for manual, flexible automation, or hard automation techniques. The manual cost is constant and is usually associated with piece-rate systems. The flexible automation system's cost per assembly drops linearly until the capacity of a single system (machine) is reached. The cost line exhibits a step function through the capacities of adjacent systems. However, a smooth curve can be used to approximate the varying cost per unit. Hard automation again shows a linear reduction in cost per assembly. Because the capacity of this third option is in excess of the other systems, the cost line is linear across the figure. It will, however, become nonlinear when the capacity of the single machine is reached.

to assemble into products or subassemblies, which are themselves removed in turn by automatic devices from the work station. These mechanisms and the philosophies behind them are discussed in detail in Section 4.1.

4.1. Parts Presentation and Orientation

At a work station components are manipulated and fitted to other components to form assemblies. The easiest form of component presentation is that of bulk deposit, in which no attempt is made to separate any one component from the bulk and orientation and attitude are random. At the other end of the spectrum, components can be automatically sorted from bulk and presented singly at the pickup station in the correct attitude and orientation for direct assembly. While there are many forms of component feeders, they fall into a number of "families." Likewise, there are a few basic escapement mechanisms that permit separation and dispensing of individual items from an ordered row or column. Yet there are a plethora of shapes and sizes of these devices that have been developed for particular applications. In any automated system, it is necessary to know whether or not a component is present and correct for pickup. Hence a wide variety of sensors operated through a number of physical phenomena are available for use in feeding equipment. Components can be presented to a work station in five basic ways:

1. They may be placed in a bin or pile that the (manual) operator can take from as necessary;
2. They may be presented one at a time from a bulk holder to a pickup point;
3. They may be supplied on a reeled tape or strip for automatic or manual removal;
4. They may be manufactured as desired at the work station; or
5. They may be presented singly from a gravity magazine, feeder, or tray.

4.1.1. Bulk Feeders

This group of parts feeders present the components in a jumbled heap, and it is then up to the operator to select individual components from the heap as demand requires. One of the major problems with this kind of feeding is that the components can become tangled and jumbled together so that separation into individual parts can be problematic (see Chapter 8). Bulk feeders can also be used to supply vibratory bowl feeders with components that are then orientated and presented by that device. Bulk feeders can be classified into four groups:

1. Components presented in a tray or bin to the operator—pre-packaged kits, for instance—where the components can be separated by type, but are jumbled within specific component "pots." Another form is where a batch of components is presented

in a bin with the contents jumbled within the confines of the bin. In both cases selection of individual components is up to the operator.

2. Components presented singly to the operator by means of a conveyor. In this case, their orientation and attitude are not assured, so again the operator has to arrange the component correctly so that it can be assembled.

3. Elevating belt feeders, where components are continuously moved to a storage area at the work station, from where the operator selects, orientates, and uses the components as required.

4. Gravity feed hoppers, which release a number of components into a container or "sunken sweep area" next to the operator (from the bulk supply in the hopper). Many of these units are fitted with sensors so that the number released each time is limited. Yet again it is up to the operator to select, orientate, and assemble.

4.1.2. Orientating Feeders

The function of these types of component feeders is to regiment a mass of jumbled, randomly orientated components that are dumped in their midst into a single or multiple presentation file of components in the desired attitude and orientation for direct assembly or transfer into an assembly device. There are five basic types of these feeders:

1. *Vibratory bowl feeders* are the most common and most comprehensive of these feeders. They are manufactured in various sizes with either cast or fabricated bowls. The bowls can be designated to incorporate several features that will aid the orientation process, and in addition they can be lined to reduce noise and improve conveying action as well as reducing damage to components. The individual components are separated from the mass by vibration around a spiral track that "lifts" the components from the base of the bowl. Tooling selectors are fitted either inside or outside the bowl to orientate the components and to arrange them in the desired attitude. Components that are not correctly arranged are ejected from the tracks back into the bowl. An advantage with bowl feeders is that the tooling is an integral part of the bowl and so vibrates with it, causing the components to be under drive at every stage from the bottom of the bowl to the pickup station.

2. *In-line vibratory feeders* consist of two vibratory conveyors, one acts as a reservoir and supplies the randomly oriented parts to the second conveyor which is tooled to discharge a single line of correctly oriented parts. Rejected components slide back onto a recirculation surface that deposits them back into the reservoir.

3. *Rotary feeders* are suitable for handling screws, rivets, bushes, and other parts that have relatively simple configurations. A rotating ring, sometimes having slots machined in it to suit the component's profile, carries the components up a gravity track which is tooled to allow only properly positioned parts to be discharged. The tumbling action of the parts against the rotating outer ring can damage fragile components, so use of this unit is somewhat limited.

4. *Oscillating feeders* are also used for feeding fasteners, such as rivets and screws. In this case, an oscillating vane or similar device deposits the parts on a gravity track within the bin. Correctly positioned parts slide through a gate, which rejects those units that are incorrectly positioned.

5. *Centrifugal feeders* are a totally different type of device, in that they use centrifugal force to push the components to the periphery of a rotating disk and through the orientation tooling. The feeders consist of a rotating disk contained within a drum and usually set at an angle to the horizontal. As the disk revolves, the components are separated from the mass by either a simple friction surface or by a positive drive through cleats or pockets. After separation, correctly orientated components are allowed to slide away into specially formed discharged tracks, while the rejected parts return to the mass.

4.1.3. Reel Feeders

These feeding devices are used for very fragile "metal" parts and electronic components. There are two fundamental types. The first is formed from parts that are manufactured in a progressive press tool, for example pins for electrical connectors. The processed components are kept in the strip form, which is rolled up into a reel for presentation. The ribbon is then fed to the assembly equipment where each part is sheared from the "ribbon" as desired and assembled. The second type of reel feeder or bandolier is used for dispensing discrete electronic components into, say, an automatic board populator. In this technique, individual electronic components are mounted on continuous strips of adhesive tape, in a form where they are centered across the tape and are at standard pitches along it. The tape consists of either just one type of component *or* it can hold different components arranged in the sequence that they are to be placed into the assemblies. It is important that the individual components are not damaged or distorted in any way since this would preclude assembly from the tape into the printed circuit board.

4.1.4. Fabrication on Demand

There are a number of components that do not lend themselves to any type of automated feeding or manual sorting or handling. Examples are springs that get entangled, thin components that do not maintain a rigid shape, and components with highly complex shapes. In many of these cases, individual manufacture on demand at the work station is the easiest solution to the presentation problem. The manufacture of parts at the point of service should always be considered even if the part can be supplied by an automatic feeder. The economics of the alternatives should be evaluated before commitment to either option. This is particularly true when there are dedicated machines that perform the fabrication, such as spring coiling machines, forming machines and gasket dispensing machines.

4.1.5. Gravity Magazines and Stacks

It is not always economical to feed and/or orient components automatically. However, assembly related cost savings can be achieved through the use of manually packaged dispensers in which components have been pre-orientated and are offered individually to the assembly station by gravity.

Long rods, thin stampings, and parts having no usable physical features for orientation can often be stored and discharged reliably from trays and magazines. Standard units are available for cylindrical rods, but for most other shapes custom designed equipment is required. In its simplest form a pile of parts is usually large in area relative to its thickness. Therefore they can be nested by rods or other guides so that the parts will slide down by gravity as the lowest component in the pile is removed from the bottom. For flat components, removal at the bottom of the pile is achieved by a blade that pushes them out from under the pile. However, where parts are nested, stripping fingers or escapements are needed to separate the parts and release them one at a time into the exit chute.

Many parts, because of surface finish or coatings, do not permit stacking or feeding methods in which they could be marred. The answer is for these items to be presented individually by indexing mechanisms that contain pockets for each item. The transfer equipment services all necessary work stations, and the empty pockets are refilled at some local loading station. In many cases, the "pockets" are removable so that different components can be delivered in their own shaped pockets either in batch or sequential arrangement. Another advantage of these specially designed pockets is that since they are fitted to conveyors, this permits the presentation of discrete components at a work station in a pre-orientated and correct attitude that allows for direct transfer to the assembly unit by means of a simple pick and place unit.

4.1.6. Ancillary Feeding Equipment

The majority of the feeding devices previously discussed do not by themselves present separate individual components from an ordered row or column for pickup or transfer, neither do they check for the presence or nonpresence of a component at the escapement device. Similarly, the actual assembly task is performed by devices other than that directly responsible for the feeding and presentation of the component. These three pieces of essential equipment are discussed in the following sections.

4.1.7. Escapement Devices

While escapement systems take many different shapes and configurations, there are in fact only six different types:

1. Rachets, which are oscillating levers that feed components singly.
2. Shuttles, which are slides that operate at 90° to the conveying track and push

one component out of the end or bottom of a track for each movement of the device.

3. Drums with cutouts around the periphery profiled to suit the components, which are collected singly in the cutouts and deposited on the delivery track or at the station.

4. Wheels, which like the drums have profiled apertures into which single parts are collected for dispensing onto a delivery track.

5. Gates, which are the most common form of escapement and take the form of a simple lever that opens and shuts to allow parts to pass one at a time.

6. Jaws, which are usually spring loaded and fitted at the bottom of delivery tracks to stop the component in a known position and stop it falling through the bottom of the track. When needed, another device pulls the part through the jaws.

4.1.8. Sensors

Sensors are used to indicate the presence of a component at the pickup point, as well as indicating when the number of components in the bottom of a feeder is becoming low and replenishment is required. They are also used to indicate malfunctions through their "nondetection" of parts. There are five basic types of sensors as described below:

1. Electromechanical limit switches, in which a switch state is changed by contact with a component. The sensitivity of these devices is quite wide, although it has been known for the sensors themselves to prevent a very light weight component from passing, due to the spring force of the actuator.

2. Air-electric devices cause the switch state to be changed when a regulated airflow is restricted by a passing part.

3. Optoelectronics use interrupted light beams or reflected light beams to change the state of a switch. The spectrum of the light beam can be chosen to suit the application.

4. Proximity sensors use inductive or capacitance fluxes to indicate the presence of an object within their field of "view." They are not limited to metallic parts, but can detect the presence of components made from almost any material, provided it is within range.

5. Low-voltage circuits that are completed when a metallic part enters the local area and "closes" the circuit.

The use of vision systems has been deliberately excluded from this section, since it is considered that they are too sophisticated (and expensive) to be used for a simple two-state switching function. Vision systems are discussed in detail in Section 8.3.

4.1.9. Placement Devices

The majority of part processing operations require that components be transferred from the pickup point (escapement) and taken to another point on the work station for direct assembly into another component or subassembly. This transfer must in no way distort the orientation and attitude of the component, albeit sometimes the transfer device will deliberately rotate the component about one of its planes, so that the component is presented correctly to the assembly device. This will only occur when the assembled attitude and orientation of the part is different from that which is practical for automatic feeding. For instance, a headed pin may be fitted horizontally, but the best method of feeding is to have it vertical and "hanging" on the underside of its head.

The devices used for placement vary from simple shuttle mechanisms, through non-servo pick-and-place manipulators, to fully servoed robots. The choice is controlled by the sophistication of the task, the speed of operation, the flexibility of the system, and the allowable cost of the work station. The factors that ultimately determine the configuration of the placement device are those of the assembly task which defines the movements, component structure that defines the gripper construction, and the fastening technique required, which defines the strength and sophistication of the "placement device." These factors are discussed in the following sections.

4.2. Assembly Tasks

With assembly related tasks occupying 53% of the throughput time and 22% of the total labor cost incurred in a "typical" product, it is worth examining what constitutes these tasks which are so influential on the prime cost of any product. Every assembly task is simplistically a pick-and-place motion, where one part is collected and taken to a second, to which it is fitted. Within this "simple" action, there are however seven distinct and different actions involved:

1. *Position.* This is the movement of the assembly device or placement unit to a particular attitude or position in space. For instance, the gate within the assembly device will open to accept a component, or the pick-and-place device will move to the pickup station of the bowl feeder and wait there with its jaws open.

2. *Grip.* This is the closing movement of the device when it receives a signal (or after a prescribed time) that it has arrived at the correct place and that a component is there to be grasped. The actual force exerted during this operation can be very critical if the component being collected is very fragile or delicate. If needed, the gripping action can be monitored to ensure that over loading does not occur.

3. *Pick.* This is the action that transfers the component from its previous location to a new location within the transfer device. This operation can again be monitored

to check that transfer is achieved, since if only partial transfer occurs, which leaves the component partially out of its original location, then damage could be caused to system if this malfunction was not "observed" and acted upon.

4. *Move.* This action causes the transfer mechanism to move, taking or causing the component to move to another accurate spatial position within the work station. It could at the same time cause the attitude and orientation of the component to move relative to that at pickup, such that it is correct for the actual assembly action.

5. *Place.* This is the presentation of the transferred component to the second component that is already at the assembly site. At this point, the two components are still distinct and in many cases would fall apart if removed as a pair at this stage. The orientation and attitude of the two parts to each other is important and in many cases must be closely controlled. Sometimes the presented component will pass over or into a location feature of the second component so that the orientation is assured.

6. *Fit.* This is the action that fits the two components into one assembly. This can be done by relative rotational motions of the two parts to achieve a screwed or friction welded assembly (see Section 4.4).

7. *Feedback.* When the fitting action is complete, it is often important that the system be made aware of this so that the assembly can be ejected from the station and the next assembly processed. It is also necessary for management to know the number of acceptable assemblies produced. Consequently, the assembly can be tested prior to ejection and the information assimilated. Feedback can come from internal or external sources—the former check that what should have happened during the assembly did in fact happen, whereas the external checks are to ensure that, for instance, the ejected unit did not leave the station, or that the new components are ready for processing.

The ease or complexity involved in handling and processing an assembly is very dependent upon the structure, material, and size of the components used. Evaluation of some of the types of components that could be encountered and their problems is worthwhile.

4.3. Component Variability

If a "typical" electromechanical product is evaluated in terms of the materials used for its construction, then handling and assembly problems related to the material and not the component can be identified. In general, it can be said that such a typical product uses plastics, metals, printed circuit boards, electrical components, electronic components, wiring, and "unclassified" components within its construction. These materials are evaluated below:

Plastic. This is one of the most common materials to be found in electro-mechanical products. One of the major problems is the possibility of crushing or marring it, while it is being handled or processed. The product's appearance can also be affected by variation in the quality of the plastic or the process. The faults here manifest themselves in flash, distortion, or discoloration. If the flash or distortion is severe, handling and actual assembly may be limited if not impossible.

Metal. This material is used for plates, springs, screws, pivots, etc., all of which require sophisticated handling techniques, since the majority require to be accurately positioned and inserted into other elements of the unit.

Printed circuit boards. In one form or another, these will be in every electro-mechanical product. There are a number of specialized machines available for automatic population of PCBs. Because of the usual density of components on PCBs and because the tracks occupy the majority of the surface(s) of the board, they can usually only be picked up by their edges.

Electrical components. These are usually very dense components such as transformers and solenoids. The major problem is in handling and accurately positioning small, but relatively heavy parts. The actual mode of handling requires that they be held firmly, but without a force that could cause distortion or damage, and in a manner that prevents the component twisting when performing the actual fit of one electrical component to another.

Electronic components. These usually enter the system on bandoliers, in special plastic tubes or loose. There are a number of specialized machines that will process components supplied either on bandoliers or in tubes. Loose items cause problems and they should be avoided at all costs. It is usually possible to develop a system for handling these components, but it may be more cost effective to use semi-automatic or manual techniques to achieve the desired result.

Wiring. Looms are common to most electro-mechanical products, but the methods of connection vary. Due to the flexible nature of the looms, they create a number of problems for automated assembly. The end connections can be plugs, soldered, bolted and many other options, all of which require accurate positioning of the wire end. This accurate gripping and positioning presents an almost impossible problem to solve by cost effective automation, and perhaps the best option is to use semi-automatic or manual techniques.

Unclassified components. These are items that do not fit into any of the other six groups, such as foams, shrink tubing, adhesives, etc. These items are usually associated with specific problems and products and often have their own unique handling equipment. If not, then economics will determine whether the choice is special purpose device, semi-automation, or manual assembly.

The material, configuration, and function of a product is an indication of the choice of assembly method that may be used to secure it to another part. The next section looks in depth at the various fixing and forming techniques that are available to any FAS.

4.4. Fixing and Forming Techniques

The primary tasks within the manufacturing industry are those of the fabrication and shaping of materials, followed by their joining and assembly into products. Fixing and forming techniques are therefore all pervasive, and there can be very few areas of the manufacturing industry that do not make use of at least one joining process. The various processes for fixing and forming of materials are very diverse in their applications. Welding for instance, can be used for the joining of large pressure vessels or the manufacture of miniature electronic components. Similarily, mechanical fasteners are used in the construction industry (bridges) as well as in the precision instrumentation industry. Adhesive bonding is presently used in a number of limited applications, but with the movement toward lightweight materials, honeycombs, and plastics, the use of structural adhesives will be used more and more, especially in place of fasteners.

4.4.1. Fastener Groups

There are two distinct groups of fastener methodologies—those that result in a permanent joint that requires destruction of one of the elements in order to achieve separation, and semipermanent joints whereby the elements can be separated at will, without damage, assuming that the correct tools are used. Examples of each group are:

Permanent	Semipermanent
Adhesive	Bolting
Crimping	Clipping
Riveting	Pinning
Soldering/Brazing	Screwing
Staking	
Stapling	
Welding	
Arc	
Friction	
Laser	
Spot	
Ultrasonic	

One point to be noted is that three of the semipermanent methodologies each require a hole to be drilled and/or tapped, punched, or accurately reamed. These expensively produced holes are then filled with an expensive fastening device. It would seem that one way of reducing the costs that are allocated to assembly tasks would be to design out the requirements for all but the most essential screws, bolts, and dowels.

4.4.1.1. Permanent Fasteners.   Adhesive. This fastening process is used more and more since it enables components to be joined together without the requirement of drilling holes that are then filled with a screw or other device. With the use of suitable adhesives, components of different materials can be satisfactorily joined together; also, the use of adhesives allows for the securing of "soft" materials such as foams and acoustic materials. If one assumes that the fundamental difference between adhesives and sealants is very subjective—since they both seal and stick—then the range of materials that is available is immense. There are problems with the shelf life and the curing of the "adhesive" in the feed nozzles of applicators, but the advantages of automatic dispensing and the low cost of this medium greatly outweigh the disadvantages.

Crimping. This involves the deformation of one component into a groove/hole in another component. It is peculiar to certain industries and is limited by the thickness of the component to be deformed.

Riveting. This generally refers to a one-piece fastener that is used to secure components by the deformation of its headless end. This results in a large clamping and frictional forces. There are two basic classifications of rivets:

(i) Rivets that require access by personnel and/or tooling to *both* sides of the elements to be joined together.

(ii) Rivets that can be inserted and deformed from *one* side of the elements being riveted.

The first classification is the traditional method of riveting and is applicable to a whole range of rivet types, configurations, and sizes. The second classification has been around for the last quarter of a century or so. It requires specialized tooling and can only be performed with a specific style of rivet. The technique for expanding the headless end requires that an axial tension force be applied to a special insert. This causes expansion of the headless end until the desired clamping force is reached. At this stage, the special insert shears, leaving one portion of the insert on the "inside" of the riveted joint. Automatic riveting guns are standard industrial equipment, and the use of robotic riveters is being evaluated under the USAF's ICAM project.

Soldering/brazing. These are similar processes in which two elements are joined via a "filler" metal. The filler metal in the molten state penetrates the interface void between the two elements by means of capillary flow, and when it has solidified, a joint has been effected without the melting of either of the elements being secured. The simplistic difference between soldering and brazing is the melting point of the appropriate filler metal, with that for soldering being below 800°F and that for brazing being above 800°F. There are a number of automatic machines whereby both brazing and soldering can be done within an automated mode. Equipment is available as standard for flow soldering printed circuit boards, so this joining process is compatible with the work ethics of the FAS.

Staking. This generally refers to the use of lanced tabs on one component being inserted into slots in another component and then being deformed such that a permanent joint results. This technique primarily belongs with the sheet metal industry, although the concept should not be ignored by other industries.

Stapling. This technique and stitching are used under the following circumstances:

(i) when it is necessary to join certain dissimilar metals, or nonmetallic materials to metals;
(ii) when high-speed fastening is required;
(iii) when elimination of pre-drilled or pre-punched holes is desired;
(iv) when it is desired to fasten together coated materials, with minimum disturbance to the coating.

Arc welding. This method of joining components together uses high energy and a filler metal. It is used in a lot of automated systems especially with robots.

Friction welding. This is solid state welding that is accomplished by the relative rotation of the components to be joined. This results in frictional heat that causes plastic flow, and the two components are forged together with an axial force.

Laser welding. This is the use of a laser to heat a local area on a component such that a molten zone is formed, which then results in a solid state weld.

Spot welding. This method of welding applies an electrical current and a pressure to a localized area such that the components are metallurgically joined together.

Ultrasonic welding. This welding technique applies to both metals and plastics. It involves the application of ultrasonic vibrations to the components being joined such that solid state joints are made.

4.4.1.2. Semi-Permanent Fasteners. Bolting. This term is usually associated with the usage of large-diameter bolts (bigger than 6.00 mm) that are screwed into a threaded hole (another component or a nut), whereby a clamping/frictional force is generated of sufficient magnitude so as to ensure that the two parts are held rigidly together. Similarly the fit of nuts onto studs results in a clamping force that can be used for the retention of components. The use of standard assembly machines, through which bolts and nuts are screwed into/onto other components is a well-established technique. Likewise, the requisite clamping/frictional forces are achieved via the usage of proprietary torque wrenches.

Clipping. This process utilizes the elastic properties of material such that the initial assembly of the two components requires the automatic elastic deformation of one or both units. At the state of final assembly the deformed components "snap" into recesses such that they cannot be disassembled without the application of a special tool (if at all). This fixing medium is of particular benefit to the automated assembly process in that it requires no additional components to achieve assembly and that the process is satisfactorily achieved with only a light axial force.

Pinning. This is the use of a dowel or spring pin to hold two components together. Before the "pin" can be inserted it is required that a suitable hole be machined in both components to be joined. The quality of this "hole" is dependent upon the style of pin used and can either be an accurately reamed hole or a drilled hole of wide tolerance. Another style of pinning is the use of split pins. However, as this requires a second operation to deform the legs of the pin, this option is considered inferior (for automated assembly) to those styles that require only an axial force to achieve assembly.

Screwing. Generically, screws are threaded devices that are less than 6.00 mm in diameter, and are used for joining components together. They are available in two distinct styles:

(i) Self-tapping, which requires only a drilled hole since the screw cuts its own thread as it is screwed into position.

(ii) Standard threading, which requires a tapped hole in one component, such that a clamping force can be generated between the underside of the screw-head and the top surface of the component with the clearance hole. Generally screws are not used with nuts, as this minimizes the number of (small) components used in any one assembly operation.

The automatic feeding and assembly of screws into components is of medium technology, with the screws being directed to the driving member of an automatic screw-driver by means of a feeder.

Table 2 is a matrix of material groups versus fastening techniques. The table is deliberately limited to the standard, present-day fastener styles, albeit it is acknowledged that new techniques and new applications of existing processes become available almost daily.

4.5. Work Station Definition

Before a work area can be designed, it is necessary to define the criteria for delineating the functions that each work station (that most constitute the work areas) will be capable of performing. That this is necessary is evident from the wide selection of assembly and feeding methods that are available to the designer of the work station. Obviously, not all of these options are wanted, nor could they all be used, hence a very subjective capability profile is made for the work station(s) that is product-*type* dependent.

Any product can be broken down, metaphorically speaking, into a series of self-contained subassemblies. In order that the individual work stations retain their prod-

Table 2
Material Groups vs Fastener Techniques Matrix

	Materials					
	Metals					
Fastener group	Ferrous	Nonferrous	Plastic	Ceramics	Elastomers	Papers/fibers
Adhesives	•	•	•	•	•	•
Crimping	•	•				
Riveting	•	•	•			
Soldering/brazing	•	•				
Staking	•	•				
Stapling		•				•
Arc welding	•	•				
Friction welding	•	•	•			
Laser welding	•	•				
Spot welding	•	•				
Ultrasonic welding	•	•	•			
Bolting	•	•	•			
Clipping			•			
Pinning	•	•	•			
Screwing	•	•	•			

uct independence, it is desirable that these subassemblies be determined by task as well as by function, albeit that it is necessary each be a self-contained unit that can be handled and transported without the risk of falling apart. Many product ranges use common items and subassemblies that are combined with unique components or modules to make up the individual product types. It therefore makes sense to define the subassemblies so that maximum variability of product options is maintained as far up the assembly chain as possible, i.e., do not commit a subassembly until the last possible assembly station. The other factors that define a work station's configuration are those of output rate and work content. The correct relationship between these two factors can result in the assembly line being balanced with little or zero nonproductive activities.

BRSL[2] used the same philosophy to define their work areas. They based their analysis on a single product type that consisted of four main subassemblies, which are built independently and only come together for the final assembly. This modular approach meant that completed subassemblies could be stored and sold independently if desired. Each subassembly is available with different options, and different combinations are used to cater to the various requirements of different countries and clients. For a particular order, the specific subassemblies are drawn out of the stores, customized, then assembled together and returned to the stores for dispatch. With the exception of the printed circuit board (PCB), all of the components that are used in the four subassemblies are purchased. The PCB is populated with components within the FAS, which means that the total number of major assembly tasks is six—the four subassemblies, the PCB, and the final assembly—therefore, the total assembly process can be broken down into six distinct work areas.

The assembly requirements of each work area were analyzed to identify the operations involved in assembling each component contained within each subassembly. Each requirement was then assessed to determine the fixing method and the required accuracy of placement so that the desired result was achieved. These factors were then considered to determine whether the operations could be performed automatically or whether they needed a human assembler. After all of the operations and tasks required to assemble the complete product had been examined and tabulated, it was noted that a number of the manual tasks could be done automatically, provided that minor design changes were incorporated into the product. (It had not been designed with automatic assembly in mind—see Chapter 5.) These changes involved rearrangement of screws so that they could all be inserted from the same direction, the replacement of self-tapping screws by threaded inserts, and the mechanical fixing of certain items instead of using adhesives.

There were a number of operations that because of the design and functional constraints of the product could only be performed manually. The loose wires and wiring looms had to be wrapped around or threaded into terminal posts. Their very flexible nature meant that any automated process would be very difficult to prove as being a cost effective alternative. Likewise, other flexible components, such as "coil cushions" and "prism foam," were difficult to feed and handle manually, let alone automatically. Many of the springs in the assemblies were of the type that easily entangled. Many also required preloading by a process that was not amenable to automation. Finally, many of the components, because of their design, would not fit together in a smooth and easy operation. This last feature is one of the penalties that has to be paid when a product uses a lot of different purchased standard components. It is also incurred when the product's cost structure will not stand the cost of specially designed components that would facilitate easy automatic assembly.

4.5.1. Work Station Classification

An extensive examination of the six assembly processes enables their classification in terms of fixing styles; high or low precision of location; and manual or automatic performance of the tasks. The results of these classifications are given in Tables 3 and 4.

4.5.2. Work Station Line Balancing

To determine the actual number of work areas and work stations that are required to satisfy demand and to identify the split between manual and automated

Table 3
Precision of Location and Fixing Style for Each of the Six Major Assembly Tasks

Assembly[a]	Components		Location		Fixing style					
	Types	Quantity	High	Low	Adhesive	Screw	Clip	Solder	Ultrasonic	Other
V	24	41	21	3	4	8	2	1		9
S	15	22	14	1	2	6			1	5
D	11	26	10	1			1	1	2	7
P	7	9	6	1		6	1			
F	17	30	15	2		7	3			5

[a]The printed circuit board is *not* included in this table because it is primarily an automatic operation and the fixing is by flow solder bath.
[b]Please note that "other" refers to being held in place by other means.

work stations were the reasons for BRSL[2] conducting a line balancing calculation. This calculation was based around the known demand for the product, the "work content" of the various tasks, and the number of "allowable" hours per year that the FAS is operational. The production system was balanced as follows:

Required production rate: 300,000 units per year
Assuming a 48-week year: 6,250 units per week
Assuming an 80-hour week: 78.1 units per hour
or: 1.3 units per minute

Therefore, one unit has to be produced every 46 sec at 100% efficiency. On the basis of one production line, and with consideration of the amount of work that has to be done at each work station to meet criteria in addition to that of the assembly task—for instance, the need for each stage to be in a form that precludes falling apart—it was decided that the time allowed for each work station was too short.

Table 4
Split between Manual and Automatic Performance of the Six Major Assembly Tasks

Assembly	Manual tasks	Automatic tasks
V	21	12
S	5	7
D	8	7
PCB	1	5
P	6	6
F	15	0
TOTAL	56	37

Therefore, two parallel work centers were considered, since this has the effect of doubling the processing time at each work station.

Based on a twin production system, the allowable processing time for each work station is 92 sec (1.53 min). However, this would only yield the required production rate if both work centers performed without breakdown and were trouble-free for the duration. To take account of the occasional interuption of production for any one of a variety of reasons—i.e., an operator at the washroom or ill, components arriving late at the work station, a batch of faulty components detected and being removed, etc.— a 15% contingency was included in the calculation. This meant that the work stations were "allowed" 1.53 min to perform 1.30 min of work. For the manual stations this 1.30 min was factored by another 20% to take account of operator variation, fatigue, and personal allowances.

The line balancing procedure was conducted for all of the tasks so that they were grouped in compatible "lots," each with a maximum work content of 1.30 or 1.08 min, respectively, for automatic and manual work stations. Where necessary, duplicate work stations were used so that the throughput of the paired work stations matched the required production output. The split between manual and automatic work stations is shown in Figure 15 for one of the two identical work centers. There is, in addition, a separate manual work station at which assemblies that fail inspection tests are repaired and refurbished prior to being reentered into the production process.

The FAS as a production unit is summarized below:

Required production per year: 300,000 per line year
Required production for two lines: 150,000 per line
Number of work stations per line: 38
Number of manual work stations: 21 per line
Number of automatic work stations
    (includes dedicated equipment on the
    PCB line): 17 per line
Ratio of manual to automatic units: 55%:45%

4.6. Work Station Design

The major requirement of a work station is that it should be totally flexible in operation, physical layout, and technology. One aspect of BRSL's[2] brief was that the technological level of the work stations would be capable of the following:

- immediate practical application;
- reliable in-production usage;
- readily repairable by staff who were adequately trained but not necessarily highly qualified.

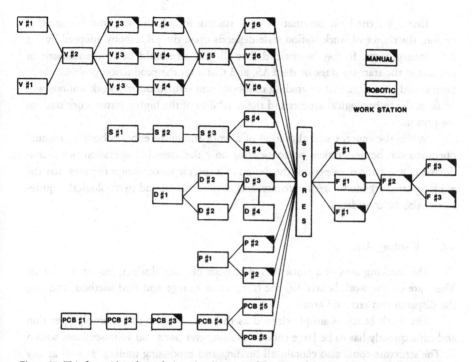

Figure 15. This figure identifies and gives the allegiance of the 38 work stations that constitute one of the BRSL designated work centers. The work stations are arranged in six work areas coded V, S, D, PCB, and F. Multiple work stations are used where the capacity of a single station is insufficient for acceptable line balancing.

There are four different basic types of work station. The first is a hard automation unit that performs a specific task. It may be fitted with removable and adjustable tooling that permits the task to be performed on differently sized or different shaped components. Its use within the FAS is limited to the performance of tasks that analysis shows are economically viable, even though the use is limited to one task on one product. Flexibility and automation are the identifiable features of the second type of work station. These consist of robotic or other adjustable systems that can quickly be reconfigured by hardware or software changes to suit various product tasks. All tasks performed by these units are performed automatically. The third choice is that of the semiautomatic work station, such as those marketed by Lanco and ISM, which make use of automatic modular handling and processing equipment to enable a human to perform the various assembly tasks. The last style is that of the manual work station, where the components are collected from bins or trays and the assembly is performed entirely by hand, although automatic hand tools such as autoscrewdrivers may be used.

Except for the hard automation work station, for which there may be no other option, the choice of work station style depends upon the philosophy adopted by the FAS management. It may be decided that all stations should be semi-automatic or manual at the start-up stage of the FAS, and that after the production problems have been sorted out, a gradual upgrading to robotic and semiautomatic work stations will occur as the technological aspects and the reliability of the higher status work stations are proven.

While the concept and allocation of a work station to either robotic or manual operation can be undertaken by considering only the assembly operation being conducted in a specific time period, the design of a specific work station requires that the working area, ergonomic requirements, and environmental and psychological requirements also be considered.

4.6.1. Working Area

The working area of a work station consists of three distinct, but related units. These are (i) the work bench, (ii) the component storage and feed method, and (iii) the dispatch and retrieval area.

The work bench is simply defined as the area used for the assembly operation and consequently has to be large enough to maneuver parts and subassemblies within it. The structure must also contain all holding and processing tooling, as well as any special purpose systems that are required. Flexibility of this unit is obtained by using a standardized form of attachment for the tooling and power systems whereby they can be quickly snapped on or off.

The more complex problem of component storage relies on criteria such as how the components are to be laid out around the work station, whether the work area is to be robotic or manual, and what type of feeding system is to be used. There are two basic types of components to be stored around the work area. First, there are the subassemblies that have operations performed on them or components added to them prior to their leaving the work station. The second group of components are those that are stored at the work station and are either fitted together to form subassemblies or are fitted to subassemblies arriving at the work station. Subassemblies are dynamic; they therefore require an area for arrival prior to processing and an area for dispatch afterwards. Piece parts on the other hand are only stored at the work area and, consequently, require only a single storage area per component. Table 5 shows the quantities of subassemblies and piece parts that BRSL[2] determined would be stored at each work station.

The space required at a work station to contain subassemblies is directly related to the size of container in which the subassembly is issued from the stores. The Conserve-a-trieve system advocated by BRSL is simplistically a "pigeon hole" system into which a standardized bin will fit. These bins will contain all of the subassemblies and piece parts that are automatically issued to the various work stations. Since it is nec-

Table 5
Number of Subassemblies and Piece Parts Stored at Work Stations for Each of the Six Major Assembly Tasks

Assembly[a]	Work station	Subassembly IN	Subassembly OUT	Piece parts
V	#1 Robotic	2	2	4
	#2 Robotic	2	1	2
	#3 Manual	1	1	5
	#4 Manual	1	1	2
	#5 Robotic	1	1	3
	#6 Manual	1	1	11
S	#1 Robotic	2	1	1
	#2 Robotic	2	1	5
	#3 Robotic	3	1	2
	#4 Manual	1	1	3
D	#1 Robotic	2	1	3
	#2 Robotic	2	1	4
	#3 Manual	2	1	1
	#4 Robotic	4	1	—
P	#1 Robotic	2	1	1
	#2 Manual	1	1	3
F	#1 Manual	1	1	9
	#2 Manual	3	1	1
	#3 Manual	2	1	6

[a]The printed circuit board is *not* included in this table because it is primarily an automatic operation and the fixing is by flow solder bath.

essary that these containers stay on the automatic conveying system so that their position and destination are known at all times to the control computer, the contents of the bins will be held in secondary containers. These secondary containers held within the bins will be in three forms:

1. Trays, that will hold preoriented parts in the horizontal plane;
2. Subbins, in which random parts will be held in small batches; and
3. Magazines, that will hold preoriented parts in the vertical plane.

It will be possible to convey large components in bins that do not have a secondary container, provided that the components can be removed (and replaced) without the bin being removed from its known position and orientation on the conveying system.

In the system analyzed by BRSL, the work stations could either be manual or robotic. For a number of reasons, the robotic stations do not contain systems that will be able to detect the orientation and attitude of randomly presented subassemblies. It is therefore necessary that all parts arriving at a robotic work station be preoriented and that this orientation be maintained throughout the entire assembly operation. Since the majority of subassemblies consist of a rectangular plastic molding with substantial thickness, it was decided that a tray system was the best method for conveying the items in the desired orientation.

The trays are manufactured in vacuum formed plastic and contain a series of purpose designed pockets that will hold a specific component or subassembly. The trays present items to the work stations in the unique orientation that is maintained throughout the operations conducted at that work station. After processing, the trays with subassembles in the correct orientation to suit the next operation are placed into empty bins. During the course of an assembly operation the tray and contents are arranged on the work area in a prescribed position. Since the number of discrete items will be less after the assembly operation, and because the outline of the module could be different, many of the trays are formed to accept several stages of subassembly, although the trays are foolproofed so that the correct orientation for the correct assembly stage is assured. The tray size is limited by the need for it to fit within the internal dimensions of the bin. However, the thickness is limited only by the component being transported, so it is possible that several trays containing the same items could be transported in one bin. This system gives a fairly accurate method of preorienting items and, apart from being relatively inexpensive, ensures that the subassemblies can be removed and replaced in the conveying system with a minimum of complication. Table 6 shows the numbers of bins, trays, subbins, and magazines that BRSL determined would be necessary to hold the various parts and subassemblies issued for the production of a single unit of the product evaluated under the study.

Each work station contains a bulk supply of piece parts that are common to many subassemblies. The method of handling and presenting these items depends upon whether the work station is robotic or manual. The variety of automatic component presentation devices has already been discussed in Section 4.1. However, BRSL determined that while several of the items serviced at a robotic work station could be handled by different methods, the use of a vibratory feeder offered the best all around solution.

The fundamental difference between robotic and manual work stations is that while it is necessary to maintain orientation of the subassembly for subsequent work at other work stations, the piece parts fitted to the subassembly at the manual work station do not have to be preoriented. This means that an inexpensive proprietary gravity feeder can be used to feed piece parts. These often consist of a tall rectangular tube with an open top into which components are poured. Individual components are made to tumble out of the feeder to a pick-up platform through a dispensing tube that is fitted at the bottom. These feeders are available in different heights and are of a modular design. Consequently, a number can be fitted together to produce a complex unit.

These gravity feeders, like the automatic feeders, will be a permanent fixture on

Table 6
Number and Types of Transportation Containers to Handle All of the Components
and Subassemblies of the "Evaluated Product"

Piece part and subassembly	Transportation method			
	Bin	Tray	Subbin	Magazine
Lid		●		
Base		●		
Snubber			●	
Screws			●	
Spring			●	
Button			●	
Hinge spring			●	
Hinge rod			●	
Coil			●	
Coil			●	
Coil			●	
Coaxial cable			●	
Shrink sleeve			●	
Magnet			●	
Lever bearing			●	
Lever				●
Lever pivot pin			●	
Lever spring			●	
Leaf spring			●	
Roller			●	
Roller bracket			●	
Roller pin			●	
PCB (V)			●	
Front cover (V)			●	
Subassembly (V)		●		
Function plate		●		
Feeder		●		
Prism			●	
First gate			●	
Second gate			●	
First gate spring				●
Second gate spring				●
Body (S)		●		
First gate solenoid		●		
Second and third gate solenoid		●		
First gate back stop			●	
PCB (S)			●	
Harness (S)			●	
Subassembly (S)		●		
Chassis (D)			●	
Base plate (D)			●	

(*Continued overleaf*)

Table 6 (Continued)

Piece part and subassembly	Transportation method			
	Bin	Tray	Subbin	Magazine
Slide 1			●	
Slide 2			●	
Slide 3			●	
Plunger subassembly			●	
Conical spring			●	
Pivot pin (D)			●	
Solenoid		●		
Tube 1		●		
Tube 2		●		
Tube 3		●		
Tube PCB		●		
Tube subassembly		●		
Chute			●	
Subassembly (D)		●		
Chassis (P)		●		
Transformer		●		
PCB (P)			●	
Subassembly (P)		●		
Bandoliered components	●			
DIL packages			●	
Odd electrical parts			●	
Enclosure case	●			
Hinge spring			●	
Hinge bearing			●	
Central PCB			●	
Wiring harness			●	
Top cover			●	
Top cover clamps			●	
Front cover			●	
Voltage cable			●	
Cable			●	
Serial plate			●	
Earth cable			●	
Completed product	●			

the work stations, and hence will be replenished in the same way, namely, by emptying subbins into their tops. Some piece parts on manual stations will be too large or of a difficult shape to be gravity fed, therefore a bench rack designed to hold a subbin can be used. As each subbin is used up, it is exchanged for a full one presented by the automatic conveying system.

Piece parts are (usually) issued from the stores in subbins within the standard bin. As the bin arrives at the docking area of a particular work station, the appropriate subbins are removed, emptied into the correct feeding device, and returned to the main bin. For this reason, all feeding devices will require some form of hopper so that parts can simply and quickly be entered into the feeder. This subbin replenishment system has the advantage that a known number of components are dispatched to any particular work station, from which an empty subbin is returned, thus eliminating the possibility of contaminated partially filled bins being returned to the automatic stores. Although magazines are usually filled by hand they can be dispatched automatically to the various work stations again through the use of the binning system.

A common operation at both manual and robotic work stations is that of screwing. This function is usually achieved by means of an automatic screwdriver, which is fitted either to the robot arm or handled by the operator. If only one size and type of screw is used within a work station, it means that an automatic screw feeding device can be used with the screwdriver. This will preorient the screws and feed them to the tip of the driver so that they can be fitted to other components by the simple action of an axial force applied to the screwdriver. The frequency of topping-up the autofeed device depends upon the usage per shift. However, due to the standard capacity of these systems, daily replenishment would be considered excessive. As with other systems that do not require frequent servicing, the position of the autofeeder on the work station will not need to be such that easy replenishment is possible by the operator without moving from the normal work area. Consequently, it could be fitted on a frame away from the direct access area of the work station.

Tables 7 and 8 show the various storage and feeding devices that BRSL considered could be fitted to both of the work stations for the processing of various components. In a number of cases there is a choice, and Table 9 classifies the devices that are allocated to specific work stations. An indication of the space occupied by these storage and presentation units is given in Table 10, which indicates the importance of good work station design so that they operate to their maximum efficiency.

While each work station is serviced by the main store and the contents of the bins are transferred automatically or manually into the working area, there is a need for equipment that conveys, positions, and presents full bins to the work stations and, when required, dispatches empty bins or bins containing processed assemblies away from that work station. The requirements of this dispatch and retrieval equipment are manifold in that it must:

(i) remove the correct bin from the main conveyor system and direct it into the correct side of the work station;
(ii) position the bin in the correct position and attitude within the working area of the work station;
(iii) when required, remove a bin (empty or containing a processed unit) from the work station and direct it back onto the main conveyor;

Table 7
Piece Part Feeding System on Manual Work Stations for the "Evaluated Product"

Piece part	Automatic	Gravity feeder	Subbin
Coil		•	
Coil		•	
Coil		•	
Coaxial cable			•
Shrink sleeve		•	•
Lever pivot pin	•	•	
Lever spring	•	•	
Leaf spring			•
Roller			•
Roller bracket		•	
Roller pin			•
PCB (V)			•
Front cover (V)			•
PCB (S)			•
Harness (S)			•
Chute			•
PCB (P)			•
Hinge link	•	•	
Hinge pin	•	•	
Hinge bearing	•	•	
Central PCB			•
Wiring harness			•
Top cover			•
Top cover clamps	•	•	
Front cover			•
Serial plate	•	•	
Earth cable			•

(iv) be controlled in such a way that the various bins, their contents, and the vacant places within the work station are known in real time to the control computer; and

(v) be adaptable to both robotic and manual operation.

4.6.2. Ergonomic Considerations

One of the advantages of automation in general and robotics in particular is that they can be specially designed to suit a given work station configuration. However, on the assumption that some work stations will be operated by humans, there are two alternatives concerning the design of the units. The first option is to design separate work stations for automated or manual assembly tasks and the second is to design the work stations for use by either machines or humans. For economic reasons and in order to retain the flexibility of the system, the second option is considered the most appropriate. Unfortunately, because Bokanovsky's process[12] is not presently viable—in the ethical sense—the work stations will have to be designed to suit the average human physique. This means that the machine using the work station format will by necessity operate at less than optimal efficiency.

A work station designed with the operator in mind should ensure that the operator's working position gives an adequate and comfortable posture. It is also necessary that the various components can be seen and reached. In addition, there should be adequate room for the operator to conduct the assembly tasks without strain. For an operator to perform at maximum efficiency, it is necessary to design the work station to suit the individual features and quirks of that operator. Since this is not a practical concept (for typical FASs) the work stations must be designed to suit the average operator.

Unfortunately, the average operator varies according to sex, race, and socio-economic environment. The work station designer can therefore be faced with the anthropometric standards for the average North American, Oriental, West European, Black, etc. Today's work force is also very mobile which means that because of transient

Table 8
Piece Part Feeding Systems on Robotic Work Stations for the "Evaluated Product"

Piece part	Magazine	In-line vibratory feeder	Horizontal belt feeder	Rotary feeder	Vibratory bowl feeder	Oscillatory feeder	Special feeder
Snubber	●	●	●		●		
Screws				●	●	●	
Spring					●		●
Button					●		
Hinge spring					●		●
Hinge rod	●	●	●		●		
Magnet (not magnetized)	●	●			●		
Lever bearing					●		
Lever	●						
Prism	●				●		
First gate	●				●		
Second gate	●				●		
First gate spring	●						
Second gate spring	●						●
First gate back stop					●		●
Slide 1					●		
Slide 2					●		
Slide 3					●		
Plunger assembly					●		
Conical spring							●
Pivot pin (D)	●	●	●		●		

Table 9
Storage Devices at Each Work Station for Each of the Six Major Assembly Tasks

Assembly*	Work station	Bins in	Bins out	Subbins	Mags	Vib. bowl	Grav. feeds	Special feeders
V	#1 Robotic	2	2			2		1
	#2 Robotic	2	2			1	1	
	#3 Manual	1	1	2			7	
	#4 Manual	1	1				4	
	#5 Robotic	1	1		1	2		
	#6 Manual	1	1	3			7	
S	#1 Robotic	2	2			1		
	#2 Robotic	2	2		2	2		
	#3 Robotic	3	3			1		
	#4 Manual	1	1	1			1	
D	#1 Robotic	2	2			3		
	#2 Robotic	2	2			1	1	1
	#3 Manual	2	2	1				
	#4 Robotic	4	4					
P	#1 Robotic	2	2					
	#2 Manual	1	1	1				
F	#1 Manual	2	2	3			4	
	#2 Manual	3	3					
	#3 Manual	2	2	4			1	

*The printed circuit board is *not* included in this table because it is primarily an automatic operation and fixing is by flow solder bath.

workers, even the usage of the "average" operator statistics for a particular geographical area can be brought into question.

In general terms, the area of a work bench is dictated by the circumscribed line defined by the maximum reach of an operator. Any components that have to be grasped should be located within this area, with large and heavy components located as close to the operator as is practical. Fine manipulation requires the dexterity of both hands and should be performed in the portion of the work area that is directly in front of the operator's body.

Operators do not just perform tasks in the horizontal plane. Work and part retrieval is often done in the space above the working surface, which means that there is a complex 3D working envelope around the frontal portion of the operator's body. The maximum working envelope for North American female and male operators is given in Tables 11 and 12. The envelope for the left hand can be taken as being the mirror image of that of the right hand, although it must be remembered that for many people, the left hand is not so comprehensive as the right for tasks requiring

Table 10
Storage Area in Terms of Units at Each Work Station for Each of the Six Major
Assembly Tasks

Assembly*	Work station	Units around working bench	Units on working bench
V	#1 Robotic	6	1
	#2 Robotic	5	1
	#3 Manual	2	9
	#4 Manual	2	4
	#5 Robotic	4	1
	#6 Manual	2	10
S	#1 Robotic	5	—
	#2 Robotic	6	2
	#3 Robotic	7	—
	#4 Manual	2	2
D	#1 Robotic	7	—
	#2 Robotic	5	2
	#3 Manual	4	1
	#4 Robotic	8	—
P	#1 Robotic	4	—
	#2 Manual	2	1
F	#1 Manual	4	7
	#2 Manual	6	—
	#3 Manual	4	5

*The printed circuit board is *not* included in this table because it is primarily an automatic operation and the fixing is by solder bath.

dexterity or judgement. There is of course a significant proportion of the population for which the reverse applies. The dimensions in the tables are expressed in terms of two coordinates, with each column representing the height above the working surface and the rows representing points along the front edge of the work bench. This means, for example, that a male can, in a plane 150 mm above the work surface, reach a point 463 mm forward of the edge of the work table and 300 mm to the right (or left) of the center line of his body. The intercept represents the furthest point along the front edge that the operator can reach on that plane.

An arrangement which will increase the available work surface easily accessible to the operator is the introduction of side benches to form an "L" or "U" shaped work station. These side benches are easily accessed by the operator through a quarter turn on a swivel chair. In general, a sit–stand work station is more desirable than either a sit or stand work station alone. If a sit–stand work unit is to be suitable for use by an operator, it must be provided with an adjustable height chair and an adjustable foot rest. To cater to the standing as well as the sitting modes it is necessary to

Table 11
Maximum Reach of Right Hand (Female)

Distance along front edge of work bench	Height above work bench surface						
	25 mm	150 mm	275 mm	400 mm	525 mm	650 mm	775 mm
0 mm (center line)	363 mm	391 mm	384 mm	386 mm	338 mm	272 mm	136 mm
75 mm (right)	375 mm	397 mm	403 mm	397 mm	348 mm	278 mm	147 mm
150 mm (right)	369 mm	394 mm	400 mm	394 mm	350 mm	273 mm	131 mm
225 mm (right)	350 mm	381 mm	389 mm	383 mm	338 mm	256 mm	91 mm
300 mm (right)	322 mm	356 mm	366 mm	356 mm	303 mm	213 mm	22 mm
375 mm (right)	283 mm	313 mm	325 mm	316 mm	253 mm	150 mm	—
450 mm (right)	213 mm	247 mm	263 mm	230 mm	175 mm	19 mm	—
525 mm (right)	97 mm	153 mm	172 mm	138 mm	38 mm	—	—
Intercept	555 mm	589 mm	597 mm	581 mm	539 mm	456 mm	316 mm

be able to adjust the height of the working surface. Typical ranges of working heights are 915 mm to 1,065 mm for females and 1,015 mm to 1,165 mm for males. The correct working height depends upon the assembly operation being performed. Most manual tasks are easily performed when the work is at elbow height. If the task requires the perception of fine visual detail, it will be necessary to raise the work piece above elbow height and bring it closer to the eyes.

Work stations where the operator will always be seated can be constructed with a lower working surface. These work stations should be provided with adjustable footrests. The distance from the floor to the top of the working surface should be 762 mm for a female and 813 mm for a male. If it is not possible to provide an adjustable foot rest, then the height of the working table should be lowered to 610 mm and 660 mm respectively.

Table 12
Maximum Reach of Right Hand (Male)

Distance along front edge of work bench	Height above work bench surface						
	25 mm	150 mm	275 mm	400 mm	525 mm	650 mm	775 mm
0 mm (center line)	450 mm	478 mm	484 mm	461 mm	405 mm	359 mm	222 mm
75 mm (right)	463 mm	488 mm	494 mm	484 mm	438 mm	372 mm	238 mm
150 mm (right)	464 mm	488 mm	494 mm	488 mm	439 mm	370 mm	234 mm
225 mm (right)	452 mm	478 mm	486 mm	481 mm	433 mm	358 mm	219 mm
300 mm (right)	431 mm	463 mm	470 mm	463 mm	414 mm	328 mm	166 mm
375 mm (right)	398 mm	434 mm	441 mm	431 mm	377 mm	288 mm	100 mm
450 mm (right)	363 mm	388 mm	397 mm	388 mm	325 mm	217 mm	—
525 mm (right)	288 mm	325 mm	334 mm	325 mm	258 mm	131 mm	—
600 mm (right)	188 mm	241 mm	253 mm	228 mm	150 mm	—	—
675 mm (right)	—	94 mm	128 mm	97 mm	—	—	—
Intercept	672 mm	703 mm	723 mm	709 mm	655 mm	575 mm	438 mm

The provision of an adequate foot rest is very important. Most work stations require them if they are to accommodate a range of normal operators. An adjustable chair by itself is not sufficient, since there are three "body elements" to be arranged, which means that of the chair, bench, and footrest, two must be adjustable. It is usually less expensive and more convenient to vary the height of the chair and footrest. If the footrest is built into the work station it should be 305 mm wide and long enough to reach across the width of the seating well. This means that it will be large enough to support the soles of both feet, it is also desirable to incline the top of the footrest, although this should not exceed 15°.

The design of the chair is just as important as that of the footrest and bench, as it forms an integral part of the work station. The use of a fixed chair can lead to poor performance in those tasks where the operator is required to work through a wide arc. The correct adjustment of the chair in relation to the footrest is important, since incorrect adjustment can cause discomfort to the operator and inefficiency for the FAS. Many studies have been published on seating based upon anthropomorphic and orthopedic considerations. The experts tend to disagree on some of the recommended chair dimensions, but the amount of disagreement is small when compared with the actual variation in the dimensions of the chairs that are commercially available. The following criteria have been found to be very acceptable for use within industry:

- The width of the chair should be at least 430 mm.
- The depth of the seat should be approximately 395 mm. Excessive variation of this dimension will cause discomfort to the users. If the backrest can be adjusted "in–out," the 395 mm becomes less critical.
- The depth of the backrest should be between 150 mm and 225 mm.
- The width of the backrest should be between 305 mm and 355 mm. Backrests wider than 355 mm may interfere with the operators elbows, and it is important that the backrests are curved to conform with the shape of the operators back. This is more important than dimension.
- The backrest should be adjustable in an "in–out" direction, so that the distance between the front edge of the seat and the front surface of the backrest can be varied between 305 mm and 430 mm.
- The backrest should be adjustable in an "up–down" direction, to account for the variation in back size and posture. A range of 150 mm will suffice for most situations.
- The seat must be adjustable in an "up–down" direction to provide for operators of different height. A minimum of 150 mm is advised.
- The seat should slope to the rear with an inclination of 3° to 5°.

Since the ideal chair probably does not exist, some compromise must be made in its selection. The chair that is selected for the operator will depend upon the nature of the tasks being considered and the costs of the units.

In some assembly operations it is required that the operator perform a search function. To aid this task, the piece part can be presented to the operator in a manner that permits easy viewing. To achieve this, the console philosophy rewards examination. This type of arrangement follows certain principles, such as placing parts in a position where they can be viewed at a relatively perpendicular angle to the eye and not require extended visual attention beyond that suggested by a fairly comfortable head position. Another principle which should be followed is that when the parts are in position for visual ease, they should also be within the work envelope of the operator. The console design can be in one of three different arrangements:

1. Flat, where a flat surface is tilted at an angle in front of the operator.
2. Sectional, where three flat surfaces are positioned in a "U" shape and are all tilted at an angle to the operator.
3. Curved, where a continuous curved surface, tilted at an angle, is "wrapped" around the operator.

4.6.3. Environmental and Psychological Considerations

The efficiency of a manual work station depends on more than the design of the work bench or the comfort of the chairs. It requires that the operator is in the right psychological mood, and also that the total environment of the work station is conducive to productive work. The three climatic elements that make up the work station environment are those of (i) illumination, (ii) noise, and (iii) heating and air conditioning.

   4.6.3.1. Illumination. There are established illumination standards for different industrial tasks. Expressed in lux these are:

- very fine assembly work—1,500
- medium assembly work—300
- sheet metal work—200

where a lux is the light given by a standard candle at a distance of 1 m from the source. Ten lux are approximately equivalent to ⅙th of a watt passing through a filament bulb or ⅟₁₅th of a watt passing through a flourescent tube. The minimum lighting intensity for areas surrounding the actual work zone should not be less than 150 lux. Additionally, there are a number of criteria that should be considered prior to finalizing the lighting on a work station.

The lighting should be focused on the task itself. Focusing on any other area of the work table adjacent to where the task is being conducted will cause fatigue in the operator and divert his attention. The problem can be overcome by installing adjustable lighting so that the lighting arrangement can be composed by the operator to suit personal and functional requirements. Too much contrast between the lighting on the bench and the lighting of the general work station could cause eye strain and may lead to work errors or even accidents because the operator may have difficulty in vision adjustment when moving from a bright area to a relatively darker area of the work station.

Discomfort and poor vision can be caused by glare. The problem will resolve itself if the lights are situated so that they do not shine into the operator's eyes and also by having the surface of the work area made of a material that does not reflect light. The visual environment can be varied by having the walls of different colors and/or providing a "distant" object to give depth perception. It is also important that the operator does not feel "shut in" at the work station, so therefore it should be arranged that a "field of view" be presented.

4.6.3.2. Noise. Noise, like illumination, can be measured, and its intensity expressed in numerical terms. The unit of measurement for noise is the decibel, and the recommended noise limits within industry are a function of the time that an operator will be exposed to a given source of noise. If the exposure time is a continuous eight hours in any one day, then the limit is set at 90 dB(A). Exposure to loud and prolonged noise will produce a limited deafness in many people, beginning with the inability to hear high notes. If the task being conducted is not mentally demanding, then it will not generally suffer if performed in a noisy environment. On the other hand if the task requires accuracy, concentration, and alertness, then exposure to undue noise will result in deterioration of the work/inspection, with scrap or rejection becoming more pronounced.

A noisy environment, apart from affecting the quality and performance of work, also has a degrading effect upon the communication system. Verbal communication within this environment is very difficult and often results in a misunderstanding of messages. Noise may be reduced by silencing the machines and/or enclosing them in insulating materials as well as facing interior walls, floors, and ceilings with sound absorbent materials. Prior to installing a work station, consideration should be given to sound-proofing all mechanical structures including the metal bins, conveyors, bowl feeders, and robots.

4.6.3.3. Heating and Air Conditioning. The ambient temperature for a work station where average tasks are being performed should be about 65°F. If the tasks are heavy, then a cooler 55°F is sometimes more appropriate. The comfort zone for the majority of people is 60° to 68°F, although in offices, where the amount of physical work is limited, the comfort zone should be raised by six Fahrenheit degrees. The real enemy is not temperature, but humidity, which requires a good air conditioning system to generate a local environment in which humans can work efficiently. The temperature environment is contained by air currents, whose velocity should be limited to about 10 m/min for optimal working efficiency of the persons within the controlled environment.

4.7. Work Station Construction

The BRSL-designed work station[2] is shown simplistically in Figures 16–19. Three sides of the work station are enclosed by double-tier conveyor units. The bins are transported on the top major conveyor until they are diverted onto the spur conveyor on the appropriate side of the work station. They are then shunted onto a lift that presents the bin in the correct attitude. The robotic work stations have additional lift systems parallel to the main conveyor which duplicate the storage facility of the gravity feeders on the manual work stations. The operation of the dispatch and retrieval equipment for the two different procedures is given below:

Procedure for bins containing trays
1. Ordered bins transferred to top spur conveyor from main conveyor.
2. Bin stops by appropriate lift unit.
3. Lift lines up with secondary conveyor (if not already there).
4. Bin is shunted onto the lift platform.
5. Lift rises to top position and tips bin toward the operator or robot.
6. When bin is empty or refilled with processed assembly, it returns to a level position.
7. Lift lowers to bottom secondary conveyor.

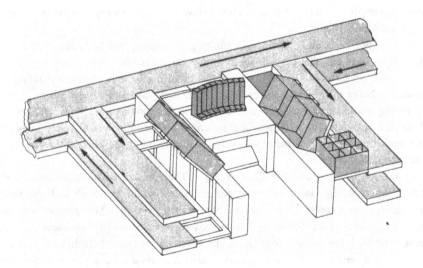

Figure 16. The manual work station is serviced by double-tier main and spur conveyors. The work bench is fitted with a presentation rack of gravity magazines and the operator can also access items from the bins presented by the vertical elevators. The elevator at the rear of the work bench exists as an integral part of the work station module. In the manual configuration it is used for emergencies only.

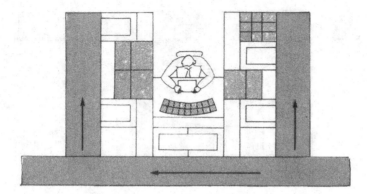

Figure 17. Birds-eye view of a manual work station.

8. Bin is shunted back onto the secondary conveyor.
9. Bin leaves work station via the main conveyor.

Procedure for bins for manual unloading

1. Ordered bins transferred to top spur conveyor from main conveyor.
2. Bin stops by appropriate lift unit.

(Continued)

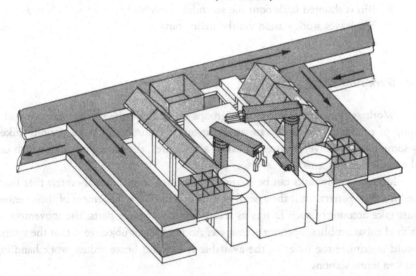

Figure 18. This robotic work station is equipped with two single-headed robots that can access items from bowl feeders or from the bins that are presented by the elevators along the spur and main double-tier conveyors.

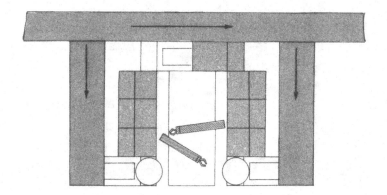

Figure 19. Birds-eye view of robotic work station.

3. Lift lines up with secondary conveyor (if not already there).
4. Bin is shunted onto the lift platform.
5. Lift rises to top position, but does not tip.
6. An audio or visual signal is activated to attract the operator's attention.
7. When bin is empty, the lift lowers it to the bottom of the bottom secondary conveyor.
8. Bin is shunted back onto the secondary conveyor.
9. Bin leaves work station via the main conveyor.

4.8. Work Area Design

Work stations are arranged into groups to facilitate the efficient assembly of a group of products. The deliberate arrangement of a number of work stations linked by some form of conveyor system is called a work area, the configuration of which can take a number of forms.

Most assembly tasks can be arranged as a series of operational centers that function in a flow pattern until the assembly is complete. The arrangement of these centers must take account of such factors as transportation of piece parts, the movement of finished subassemblies, operator access, etc. Another major objective is that the system should maximize the usage of the available space, and hence reduce work handling between work stations.

The traditional long straight-run assembly line concept is being phased out in many companies. It is wasteful of floor space, because the area between operators is allocated as being a service area to the line even though it is not occupied. The "U" arrangement is often used to consolidate line benching and it is especially applicable to high volume work requiring a relatively large "live" storage area for components. Operators can be located either on the inside or outside of the enclosed area.

Single- or double-sided spine/spur work areas also lend themselves to conveyorization, with the number of operators served per unit length of conveyor being higher than that of straight assembly lines. Piece parts can be brought forward or dispatched from the work station by the conveyor with equal convenience to all of the operators. This system can be arranged in either the herring bone or staggered configuration, with both styles offering the choice of two arrangements of seating—facing or all one way. The staggered arrangement gives each operator a section of spine, resulting in an "L" shaped work top. This reduces reaching and also the spur length for an equal area of work space. Figure 20 shows the conventional and spine/spur work area arrangements and Figure 21 shows the design adopted by BRSL. This latter arrangement shows six

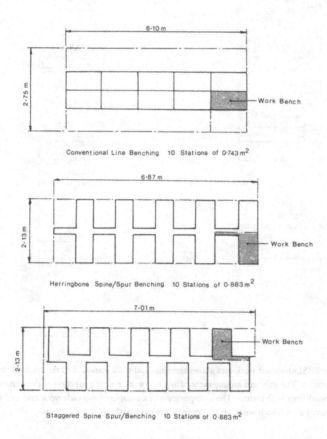

Figure 20. Three traditional work area configurations, with the extent of a single work bench shown for each version. In each type the allowable depth for the operator is the same, namely, 0.762 m. (This is the distance normal to the long edge of the work bench.)

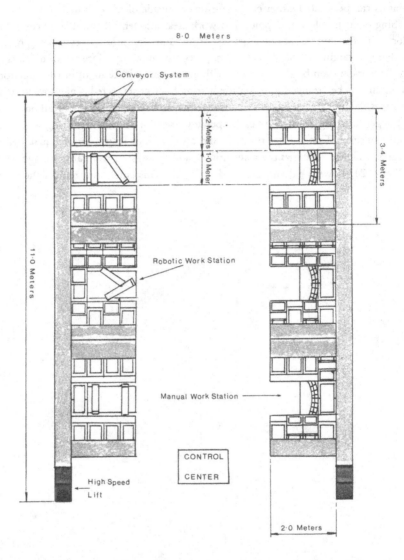

Figure 21. The BRSL-designed work area shows three manual work stations and three robotic work stations in this configuration. The mix and arrangement of any one work area is a function of the particular requirements of that work area at that time. The incorporation of a computer into each work area and the modular basis of the system allows each work area to operate autonomously.

work stations, although this number is flexible to suit the product being assembled. Work stations can be in any configuration provided that they are connected in some way to the high-speed lifts, which in turn is dictated by the position of the stores and the pitch of the aisles. The lifts have to be positioned at the ends of the aisle conveyors, with one lift servicing the two tiers of a conveyor along which bins can be dispatched to replenish work stations. The conveyors also direct the empty bins or bins containing processed assemblies to these same lifts for storage.

4.9. Computer Control of Work Areas

BRSL favored the concept of each work area regarding itself as a unit with its own successes and goals. Each work area is therefore, by design, self-contained with no direct flow of materials from one work area to another. Likewise, there is no strong dependency of any work area on any other.

The physical design of the work area suggests an autonomous unit and it naturally follows that each work area requires its own computer. It is difficult to present serious alternatives to this proposal since to place control of all local work area functions in a single monolithic factory computer would have serious drawbacks. It would need to be very powerful—hence expensive—the communications would be horrendous, and the software tortuous and difficult to test. Additionally, the FAS would be totally dependant on one piece of software and hardware. To put control in a series of small computers with no logical control computer presents unnecessary design problems.

Each work area computer will be a fairly complex system and not easy to install because of the many hardware links—to the conveyor system, to the robots, to the automatic test microprocessors, and to the factory control system. In view of this, the design should be as simple as possible and of a hierarchically arranged distributed processing nature.

The philosophy of a hierarchical system is to delegate control *down* the structure as far as possible and for each level to act as an information compressor by sending *up* only that data that is necessary. Each level should cope with local problems locally. To this end, the work area computer will deal with local hardware errors, and so on. However, it will not necessarily have to be very sophisticated software for dealing with its own problems, since whenever it encounters an unscheduled condition, it will appeal for help, either from the factory control computer or from the operator via the vdu and keyboard. To attempt to do elaborate recovery at this level would mean a very sophisticated system.

Each work area control system will be controlling a roughly similar physical configuration, with similar automatic functions. It will therefore be desirable that each system be identical in software, with any differences being a function of the user-defined parameters. The broad areas controlled by each work area computer are:

- control of local vdu terminal
- local material movement and scheduling
- communications with local work stations
- control of automatic machines
- communication with the factory control system
- security of its own system

Each work area will have a vdu through which the supervisor, or any authorized operator, can communicate with the system. A set of hardware and/or software keys will establish the identity of the user and prevent unauthorized use of the system. To save confusion, the one terminal will permit the user access to both the local system and to the factory control system. The user will not necessarily be aware of the difference. All commands such as REPORT or SCHEDULE will be simple in format and will be processed locally if they are recognized by the work area computer. If not, then the commands will be channeled directly to the main computer, which will then have the responsibility *inter alia,* for dealing with errors. The central computer's responses will similarily be channeled through to the vdu screen without any processing. In fact, the local work area computer will not recognize many of the commands since it will be responsible for getting material from the stores, stopping and starting the machines, and printing out information from the locally held data, such as the schedule, goods in hand, and shift production.

It will be possible to down load the robot training program into a work area computer and through this medium program the robot(s). There will also be a message routing system available, whereby anyone can send a message from any screen in the FAS to any other screen. The control of this will lie with the main computer and these messages will always appear on the bottom two lines of the screen. The same area may also be used for system-generated (any screen) error messages to an operator. All messages will be accompanied by bells to alert the operator to look at the terminal.

Each work area will consist of up to six work stations, though not all need be functioning. All work stations will be serviced by the two-tier conveyors via lifts at each end that connect them to the stores. The control of the lifts will be under the computer system supplied by the vendors of the stores system. The actual movement control will be by a local microcomputer, but the command to move will come from either the warehouse system or the work area computer. In either case, the work area control system has the ability to lock the lift system—to prevent a hardware foul-up—since it has the responsibility of moving the bins on and off. Again the actual movement will be micro-controlled, while the work area computer will function at the global level, issuing commands such as TAKE OFF FRONT BIN, etc.

The work area system has to coordinate with the stores system in that it has to request goods from the stores and inform the stores system when goods are to be reentered. It is envisaged that these commands will be routed up through the hierarchy, i.e., to the factory computer, and then down to the warehouse interface system and through to the warehouse control system (the interface computer having translated the demand for materials into actual locations, etc.) This may seem an unnecessarily circuituous way of going about things, but since the actual level of inventory is the responsibility of the factory control system, it seems easier to ensure that it is always aware of stock movements. Also, it seemed practical to limit the communication links between separate control systems e.g., work area–factory control, because of both hardware and software reasons. Against this is the very cogent argument that if the main factory system was inoperable, for whatever reason, then the work areas could not get goods from the stores.

The process of scheduling is also one that is ultimately the province of the factory control computer. This computer will send a copy of the schedule to the work area computer, which will use it as a scheduling basis only and not make value judgements. The factory control computer will also have the responsibility for planning and ordering parts from the stores for the automatic machines. The work area computer will always know the "stock in hand" position at the work area. At least it will know the bins received, though not their stock level. When goods are returned to the stores, they will be identified and weighed automatically on the lift system. The mode of identification will be a bar code on the outside of the bin that passes a fixed bar code reader.

Each work station will have a very simple terminal consisting of function and numeric buttons, and an LED display. Through prearranged codes, a person can request material to be delivered or removed at a given work station in a particular work area. The code will contain information that identifies the operator, which means that management information on productivity for that particular operator/work station pair is automatically noted.

Typically, each work area will contain a number of automatic machines and robotic work stations. Since these machines and work stations will be computer controlled (probably through micro's) direct communication with the work area computer will be normal. The control system of the robotic work station will naturally be more sophisticated and flexible than that of the automatic machines. It will be possible for the control system to change the programs of the robots, to start-up and stop the operation of the robots, to control the robot's operating environment by monitoring the component feeders, and, where necessary, ordering new supplies to be sent. The microcomputers will supervise and monitor the operations of the automatic machines and robots. As is necessary, they will close them down and/or advise maintenance personnel when the performance of either falls below an acceptable level.

The operational security of the work areas will be two-pronged. The first is that of philosophy in that the system needs to be of a very simple design and consist of modular replaceable hardware. The second is to make each part of the system able to work by itself, albeit only in a limited way. (This can be very useful during the start-up phase of the system.) To this end each work area computer will have its own local storage medium—floppy or hard disks—and it will be able to detect when its link with the central computer has failed. If the link fails, each work area will be able to function although on a limited basis, with the computer stacking the messages it would have sent to the disk. The main problem with a computer failure is the stores, in that the messages requesting delivery or collection of materials will not go through automatically. The answer here is that they can be displayed on vdu's and phoned through to the store's controller. While this whole procedure is necessarily stop gap and extended, it is considered that the main priority is to keep the FAS operating, even though productivity will be suboptimal while the factory computer control system is down.

4.10. Costs of Work Area and Work Stations

The cost of a work area is dependent upon the number and type of work stations that it contains. The basic costs that permit determination of the cost of a work area are:

Cost of manual work station ($)

Presentation unit	14,400
Benching	200
Conveyors	5,600
Control gear	1,000
TOTAL	$21,200

Cost of robotic work station ($)

Presentation unit	14,400
Benching/tracking	2,600
Conveyors	5,600
Control gear	1,000
SUBTOTAL	23,600
One robotic unit	20,000
TOTAL	$43,600
Two robotic units*	40,000
TOTAL	$63,600
Three robotic units*	60,000
TOTAL	$83,600

Table 13
Cost for Two Work Centers—Excluding Software Costs

Assembly	Task 1	Task 2	Task 3	Task 4	Task 5	Task 6	Conveyors	Total
V	167,000	63,600	63,600	63,600	63,600	63,600	20,000	505,200
S	63,600	83,600	83,600	42,400	—	—	10,000	283,200
D	63,600	167,200	83,600	212,000	—	—	10,000	536,400
P	63,600	42,400	—	—	—	—	—	106,000
PCB	200,000	84,000	23,800	88,000	260,000	—	10,000	665,800
F	63,600	21,200	42,400	—	—	—	10,000	127,200
								$2,233,800

Material for two work centers	$4,467,600
Engineering and design	368,000
TOTAL	$4,835,600

Cost of special purpose machines ($)

Axial lead inserter	200,000
DIL inserter	84,000
Flow solder machine	88,000
Fault finder	130,000

*This assumes collision avoidance, etc., is compensated for by the reduction in unit prices of the work stations.

From these basic costs, and using the information on work station configuration from Figure 15, BRSL[2] determined that the cost for *two* work centers would be $4,835,600 (see Table 13). It is important to note that these costs *do not* include work center software costs, which are amalgamated in the total cost summary for the FAS given in Table 23 in Chapter 10.

Product

The product is the key to the efficiency of the FAS. It is therefore important that the design of that product is in sympathy with automated assembly practices and philosophies. These philosophies are not difficult to accept or acknowledge since all of their tenets encourage simplicity, rationalization, and cost effectiveness. Any attempt to automatically assembly a product that has not been designed or modified with automation in mind will be disastrous, since all of the robots and sophisticated control equipment will not be able to compensate for the inefficiencies of that assembly's processes and procedures. Further, the system is unlikely to be cost effective, since the capital equipment will not result in improved quality or output of production.

The argument that automated assembly encourages the adoption of simple, albeit effective, procedures is supported by the fact that the intelligence level of assembly systems is grossly inferior to that of most people. The design of the product(s) must therefore be compatible with a system that has little, if any, decision making capability.

The design of a product for automated assembly requires that a systems approach be taken, where every aspect should be considered as an element within the assembly process. Each element requires that it be purchased/manufactured, handled, processed, fitted, and tested. It must also perform functionally within the operational specification of the product. Ideally, the assembly efficiency is related to the synergy of all of the individual elements that constitute the transfer of a group of inert components into a viable dynamic living product.

Products assembled within an FAS come from one of two origins. The first, and most common (at present), is a product that already exists as a manufactured item and for which a policy decision has been made to automate its assembly. This situation occurs because of a reluctance by management to spend money in designing a product—from scratch—that will be compatible with automatic assembly techniques. The reasons for this caution are manifold—the sales forecasts tend to be tempered with pragmatism, the product design is not frozen, and the marketplace acceptance has yet to be proven. The net result is that the product is designed for manual assembly with little or no anticipation of direct transfer to automated processes. Also, it is often the case that purchased components and materials have *not* been determined and specified to the *n*th degree. High-volume automated production demands this commitment if the performance values are to be satisfied at minimum cost. The transfer to automated assembly can also occur as a natural progression from a product, whose demand (initially low, but acceptable) increases beyond realistic expectations. In order to satisfy this new sales requirement, the transfer to an automated operation is inevitable if the output is to match demand, yet still retain the desired quality levels.

The second product origin is that of a totally new and conceived product, whose components and assembly procedures are not really specified, let alone frozen. The options and flexibility of the "new" product are therefore immense, with only the performance and cost per unit being the limiting criteria. There is of course, a hybrid option, that of the existing product that is to be redesigned to give perhaps better features, a more up-to-date style, and perhaps to incorporate new technology.

The existing and defined product requires several extra evaluations in order to assess the viability and/or the technical and financial implications of its automated assembly. One of the most important of these investigations is the determination of how the product is presently assembled and what are its components. It is not simply a case of looking up drawings and parts lists since the information on those documents could be out of date, incorrect, or simply not adhered to. It is also important to know what, if any, peculiar human-oriented knacks are used; whether the assembler uses selective or nonselective assembly techniques; what intelligence and animal cunning is used to achieve complete and satisfactory assembly; what is the reject rate; and what are the causes of rejects or returns from the "field." Information such as this will enable the conditions that permit/cause assembly to take place to be determined. It may be that there are some tasks or procedures used that totally preclude the use of automation without the total and complete redesign of the product.

Redesigning may be out of the question on account of the need to use up stocks of existing components and materials, or because of the requirement for commonality and interchangeability of manually and automatically assembled products. It might be the case that management might want to retain the option of manual assembly of that product, in the event that demand on the FAS's resources precludes a particular batch of products being assembled automatically. The last reason that could exclude redesign is where the product has to be approved by a body such as Underwriters Laboratories (UL), whereby redesign to suit automatic assembly could mean the expense of reapproval trials.

The design of existing products should also be evaluated to determine exactly what the assembly sequence is, rather than what is assumed; what is the possibility of components being fitted incorrectly and/or out of sequence; and what sort of actions and contortions of both the "product" and the assembler are needed in order that the product is realized. While this assessment is being conducted it makes sense to perform "value analysis" on the product and its constituent parts and processes such that the prime cost of the product can be minimized. This means checking for redundant components, changing asymmetrical parts to symmetrical parts (whenever possible) and respecifying manufacturing features of each and every component, such that they are compatible with the real requirements of the product and assembly process.

When a new product is chosen for automated assembly, then the procedure of evaluation is neater and more positive. The first requirement is to check out the function and necessity of each and every component and process recommended in the draft design and assembly specification. This "value engineering" philosophy continues with the comparison of the potential components (of the product) for similarities with other existing components already used within the FAS. It is often the case that a number of components exist, that, while different in design, perform the same function in the same dimensional and geometric fields. Adoption of an existing design *or* the use of a composite design that will satisfy the requirements for all applicable products can result in significant cost savings through inventory management and economies of scale.

Validated components are then examined in both manufacturing and handling aspects from a *machine's* point of view. This means how they can be guided, handled, manipulated; what surfaces can or cannot be touched; what is the fragility of the component; whether they can be bowl fed and how they can be presented and oriented; what are the materials; are these materials readily available; are there alternatives; do the components mar easily; and, finally, but most importantly, what quality assurance can be given to the components and materials that are procured?

The term group technology (GT) is often associated with automation. The basic philosophy is to classify components into "families" of similar requirements. As a simple example, a lot of different items known as "shafts," "wheels," "rods," "tubes," etc., could all be fitted into the same single family, which would be represented by a "composite" part, whose shape encompasses the requirements of all of the individual items that make up the family. The family would be processed by one work area (cell) with the tooling's capability profile set for the composite part. As each individual item is identified, the process performs those tasks peculiar to that component. While there are a number of coding systems that have been evolved to fully use the concept of GT, they are not essential to reap the benefits of the ideal. Often just the knowledge of the concept and the application of the fundamental principle of grouping of like parts and processes will yield advantages to the user.

5.1. Tangling and Other Problems

One of the dilemmas of automated assembly is that of automated component feeding. The advantage is that a bulk supply of components can be dumped into the bowl of a vibratory feeder, which (in the majority of cases) then presents them individually in the correct attitude and orientation at the pickup station. The potential problem with any bulk feeding is that of intercomponent tangling or binding, which means that they are not automatically separated by the feeder at a rate that satisfies demand—if they can be separated at all.

There are two remedies to the tangling of parts in an automatic feeding environment. The first is to design a specialized piece of handling equipment that will accomodate the components and ensure that the separation rate of the "bunch" will exceed that of the demand rate for the individual components at the pickup station. The alternative to this expensive development, design, and manufacture of special equipment is to redesign the components so as to preclude or at least minimize the chance of tangling, as well as to increase the rate of separation by the automated feeding device. This assumes that the components cannot be replaced by more suitable alternative items.

The most common component-related problems with automated feeding fall into five categories: (i) tangling, (ii) nesting, (iii) telescoping, (iv) shingling, and (v) orientation. Tangling is defined as the interlocking of like components when stored in bulk, such that they cannot be easily separated. Typically these components have slots, are flexible and/or have shapes that approximate hooks and rings. For instance:

1. Open coiled compression springs are the components usually associated with tangling, this being caused by having "open ends" to each spring. The solution for bulk storage is to ensure that the coil ends are closed, as shown in Figure 22. If applicable, it is preferable to manufacture the springs individually on demand at the work station. There are also proprietary spring feeders that will separate some spring types into individual items.

2. Open rings, such as retaining rings, circlips, spring washers, split bushes, etc. should, where applicable, be designed with the slot closed or overlapped or jogged, so that pressure is required to open the slot. (See Figure 22.)

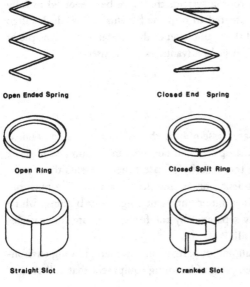

WILL TANGLE   **WON'T TANGLE**

Open Ended Spring   Closed End Spring

Open Ring   Closed Split Ring

Straight Slot   Cranked Slot

Figure 22. The left-hand versions of these simple components will tangle since they each have hooks that will engage in adjacent components. The simple solutions are given on the right-hand side. The closed-end spring is a common occurrence and the closed split ring will still permit assembly to another component through radial pressure at the joint line (e.g., key rings). The third example is perhaps more radical, but under certain conditions it would be cost effective.

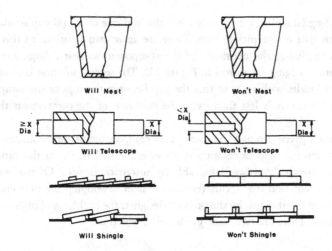

Figure 23. The left-hand versions of these simple items mean trouble to the automated assembly process, since they will nest, telescope, or shingle. The right-hand side gives the solutions to the problems with, at the top, an internal step being used to prevent the respective tapers from engaging and locking. The second illustration from the top shows how a simple change to dimensioning can prevent the studs from entering the bores of adjacent components. The lower two illustrations show how shingling can be prevented by the thickening of the contact edge so that the adjacent items will not ride over or under each other.

There are many components whose outside shape is tapered, generally because of the manufacturing process—for instance, molding, casting, or pressing—requiring that draft be added so that the part can easily be removed from the process equipment. This can lead to nesting, where adjacent components through axial pressure will (where possible) enter one another as is shown in Figure 23. The solution is relatively simple and makes use of an internal step that prevents the tapers locking *and* controls the separation distance of the components. The use of any tapered surface should be carefully examined before incorporation, since tapers can be self-locking or nonlocking. The importance as far as the feeding aspect of components is concerned is that nested components *can* be accomodated provided that they will separate under gravity. Nesting can also occur with nontapered items, which because of their inherent flexibility will, if possible, snap into adjacent parts under a given axial pressure.

Telescoping refers to components that will easily enter and leave adjacent components. Figure 23 shows the problem clearly, in that it is not the fact that the parts could get jammed together, but that the foremost one is prevented from being manipulated in any plane, except axial and roll. Hence the components get jammed up in the orientation mechanism. Again the answer is relatively simple, with all potential apertures being designed to be smaller than the potential studs.

The vibratory action and resultant axial forces of the bowl feeder on components with thin flat surfaces, results in a tendancy for these parts to climb over each other during feeding (shingling), especially when the feed rate to the pickup station is much greater than that of the pickoff rate. There are again two solutions to this problem: the first is to change the perimeter of the component so that a large step prevents climbing and wedging (as shown in Figure 23). The second solution is to arrange the bowl feeder guides with tops, so that the gap between the tops of the ramps and the bottom of the covers is less than twice the thickness of the components that would overlap and wedge.

Within a given product there are many components of varied shapes and sizes. The correct attitude and orientation of these individual items, at the time of their assembly, is invariably always achievable by automatic devices. Of the two requirements, it is orientation that incurs the most cost in developing the presentation unit, yet with a little forethought by the product designer, the problems of orientation could be minimized. There are two basic guide rules, which are:

1. Make the component symmetrical in the plane of orientation, i.e., a ball bearing, plain cube, or cylinder are always oriented.
2. If the component is asymmetrical, then make it obviously so by an exaggerated feature that permits gravity to select or reject incorrectly presented parts, or by the use of lugs that prevent access in one plane.

While a sphere is always oriented, often there is encountered the case of a sphere "modified" by means of a central hole or other slight feature. Orientation is then achieved by exaggerating the effect of the modification, so that the dimensional discrepancy is obvious. In the case of the drilled hole this generates two flats on opposite surfaces of the "sphere." If these flats are made large enough, then the component can become stable when resting on them. Other orientations would cause the part to roll or preclude entry into an orifice, etc. (See also Figure 24.)

There are many components that require orientation and that do not exhibit any geometric or dimensional features that allow this orientation to be easily achieved. (The exclusion of the vision systems from this statement is deliberate.) The answer is to add a nonfunctional feature to the component, which has little effect on the cost of the item and does not effect the performance of the component within the product, yet enables orientation to be achieved. An example is given in Figure 24.

Many flat components are identical on both sides; for example, washers that can be assembled either side up. However, there are some "flat" items that are not the same on both sides; for instance, bimetal plates, or shaped components of a single material that require orientation in the "flat" plane. With nonautomatic assembly the differences are either blatantly obvious or can be color coded to show the "correct" side. There are two solutions to this problem. The first is to make the components identical on both sides and the second is to shape the component so that it can only be presented in the correct plane and orientation. (See Figure 24.)

Many headed components are used in assemblies. These items, which are usually

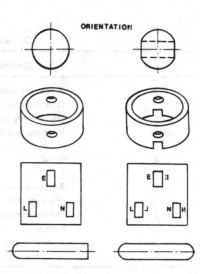

ORIENTATION

Figure 24. This figure shows a number of orientation problems. Again (with the exception of the top illustration) the problem conditions are shown on the left and the solutions are on the right. The top illustration shows how a sphere is always oriented, and how a modified sphere can be oriented by exaggerating the asymmetric feature. The second illustration show a ring with a non-functional slot added so that the cross holes can be oriented by means of a slide rail that engages in the open slot. The third illustration shows how a flat asymmetric item can be made identical on both sides through double lettering so that it does not matter which way up it is presented. The bottom illustration shows how a difficult to orient component can be made symmetrical through the use of identical end shapes.

screws, rivets, etc., are fed by tubes or between rails. In order that the components cannot fall through the rails or tumble in the tube, it is necessary that the length of the component be greater than its "head diameter" and that the component surface that rests on the feed rails be sufficiently large enough to prevent the item from falling through the rails due to rotation on the rail or lateral motion between the rails.

5.2. Product Analysis

It was stated earlier that a systems approach to product design was necessary for the best results. This means that overall consideration of the benefits, costs, and other implications of the components of the product—materials, testing routines, packaging, and so on—must be conducted. Systems approach also means knowing what is happening, when and where, and having the information in a quantitative form. It follows that if the "product" is evaluated in terms of task, component weight, assembly function, etc., the information could yield data that might result in a better product through redesign of some of its components and/or processes.

In 1976, Kondolen,[13] conducted a study so that assembly-related tasks could be defined in a systematic manner. Through stripping and rebuilding a number of different products and their components, a number of manufacturing tasks were identified that were not product specific. These tasks are shown in Table 14.

Through the use of some or all of these tasks, the majority of assembly processes can be described. Likewise, the number of products that use assembly tasks not classified within Table 14 are very few and far between. Of equal importance to the tasks themselves, is the attitude (to a given datum) at which they are performed, since this

Table 14
Categories of Assembly Tasks

Code	Description	Notes
A	Simple peg insertion	Location peg (slide/push fit)
B	Push and twist	Bayonet lock (quarter turn)
C	Multiple peg insertion	Electronic components
D	Insert peg and retainer	One component keyed by another (usually in a cross direction)
E	Screw	
F	Force fit	
G	Remove location peg	Reverse of A
H	Flip parts over	Between operation reorientation
I	Provide temporary support	
J	Crimp	
K	Remove temporary support	
L	Weld or solder	

*Typical assembly related tasks were identified by taking apart and rebuilding a number of products and their components. Included in the analysis were a refrigerator compressor, an electric saw, an electric motor, a toaster-oven, a bicycle brake, and an automobile alternator. All of the components and products could be assembled through various combinations of the above listed functions (after Kondoleon).

information is vitally important when determining the degrees of freedom that the assembly machine/robot/assembler must have in order to fulfill the tasks. Figure 25 defines the three principle directions of attachment—down, up, and sideways—and links the type of task with the direction of action. As can be seen, it is the "simple peg in hole" followed by the screwing tasks that are the most prevalent, with all other tasks way behind. The most popular direction is downwards. Surprisingly, the upwards motion is the next most frequent direction. Also, with the upwards motion, it is the screwing tasks that are preferred to the peg in hole tasks.

BRSL[2] conducted an analysis of the range of component weights that might be encountered during the assembly of an electro-mechanical product. As can be seen from Figure 11, 75% of the items weigh less than 0.1 kg. This can be compared with statements that 90% of the components used in the assembly of an automobile weigh less than 2.0 kg each and that John Deere Inc., who manufacture large agricultural machinery and earth moving equipment, determined that 80% of their components each weighed less than 4.0 kg.

The product design has also an effect upon the throughput time for the assembly process. The assembly sequence should be checked so that the expended time can be optimized, with consideration being given to changing or removing components in order to achieve the desired rate of production. An example of this philosophy is the replacement of a loose screw and loose washer by a screw and captive washer, respectively. The net result is that only one part needs to be fitted and *two* other actions are

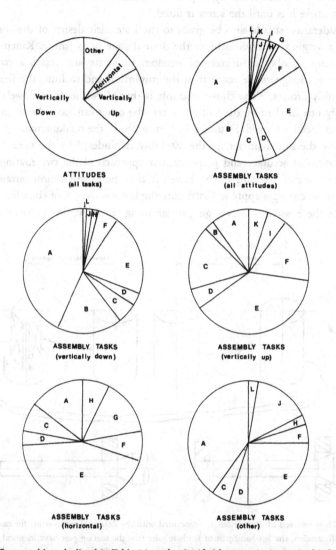

Figure 25. The assembly tasks listed in Table 14 can be classified by two criteria—attitude of task performance and popularity of use. The top left-hand illustration shows the relationship between the four categories of attitude, namely, vertically up, vertically down, horozintal (from any direction), and other.

The top right-hand pie diagram shows the popularity rating of the 12 assembly tasks for all attitudes. It can be seen that "peg in hole" and "screwing" outnumber all of the assembly tasks by a very large factor.

The relationship between a particular attitude and the assembly tasks are given in the lower four pie diagrams. (After Kondoleon.)

avoided. The first action is having to place the washer and the second is ensuring that it remains where it is until the screw is fitted.

Consideration should also be given to the kinematic design of the components such that a single action will achieve the desired state of assembly. Kinematics uses the minimum number of degrees of freedom to locate and retain a component, whereas many designers use more than the minimum and so limit the flexibility of the "assembly process." The classic example is that of two location dowels. In kinematics only one need to be the full diameter, the other can have the majority of its location perimeter removed, as shown in Figure 26. In the nonkinematic design both dowels have the full diameter, so the assembly is hindered by the need for precise relative positional accuracy and perpendicular approach of the two mating parts. If this is not done then wedging of the dowels in their holes will almost certainly occur.

The same concepts apply to shafts entering holes—the use of chamfers and radii will reduce the possibility of wedging or jamming, as shown in Figures 47 and 48.

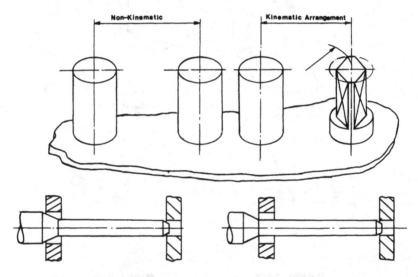

Figure 26. These two sets of illustrations are concerned with the design of components for easy assembly. In the top illustration, the left-hand pair of studs require that the mating part have identical centers and that the assembly action be perpendicular if wedging is not to occur. The right-hand pair show the kinematic arrangement, whereby one full-diameter stud restrains the mating part in the two horizontal planes. The second stud has only to stop the mating part from rotating about the full-diameter stud, hence the majority of its location perimeter can be relieved. This cutting away of the second stud means that the centers of the mating part do not have to be identical to those of the item fitted with the studs The permissible variation on center distances is a function of the stud to hole clearance, the length of contact arc, and the diameter of the modified stud. It is, however, several times the nominal clearance of the assembly.

The second set of illustrations show the undesirable condition (left) of attempting to fit the two diameters of a stepped shaft into their respective housings simultaneously. The preferred condition (right) allows one to be assembled prior to the other diameter being engaged.

Threaded shafts should have conical or rounded ends to ensure easy and correct engagement of the internal threads. Square-ended threads will not self-center and chamfered points will only self-center if the presentation is accurate enough. If a shaft is stepped and engages in two separate bores (e.g., bearings), then the design should permit the location of one diameter first, followed by the second diameter. Simultaneous location of two diameters in the same plane is not a practical automatic assembly proposition. (See Figure 26.)

When evaluating the assembly sequence of the product, it is important to ensure that (if several assembly stations are needed for a given product) the subassembly resulting from each work station can be handled using reasonable care, without the risk that they will collapse or suffer damage. It is also necessary that the relative orientation of the subassembly being "built" and the components being added to it is controlled. If this is not achieved, to the desired accuracy, then the automated assembly of close fitting parts will be extremely difficult or impossible, due to collisions between the individual components as the assembly task is attempted. This is different in cause and response to the conditions dealt with by compliance devices. (See Chapter 8.)

Many proposals for automated assembly of products make use of component drawings, sample components, or even handbuilt parts for their evaluation. This is an extremely dangerous practice since there is often a gross discrepancy between the drawing information and the components that are made and used for assembly under operating conditions. Apart from the normal gaussian distribution of dimensions within the tolerance band, there are often component batches which while not of specification are deemed to be acceptable by the inspectors or management. Human assemblers can accomodate out of limit components and will use selective assembly techniques to achieve the finished product(s). Machines will not and cannot (as yet) do selective assembly, except on a very limited and specialized basis.

Component tolerances themselves can cause problems relating to handling, feeding, and assembly processes. If they are "wide open," then precise control of orientation and attitude is limited since the guidance and presentation mechanisms must be able to accept the extremes of the dimension(s), which means that some components will slop between the guide rails, while others will be properly contained. Very tight tolerances require that the handling and feeding equipment must be at least as precise if the desired assembly is to take place without problems. Determination of the correct tolerances is a compromise between what would be preferred and the cost and technical implications of realizing that choice. In the end it comes down to optimizing the costs of product performance, manufacturing cost, and technical capability. (See Figure 27.)

Product design should take into account the importance of inspecting the quality of the product while it is still being assembled. The view that quality should be built-in and not inspected into the products is valid. However, it is still desirable that inspection functions be retained within the FAS. Inspection occurs at three levels. The first is in the goods inwards area where inspection is based upon the vendors' reliability

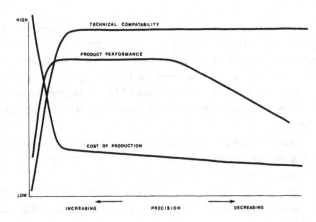

Figure 27. This illustration compares the three criteria of (i) product performance, (ii) cost of production, and (iii) technical compatibility for manufacture against the level of precision specified for the product. Where the precision is very low, the cost is low and the technical capability high. However, because of the probability of excessive wear and stressing through poor control of mating parts, the product performance will be low.

At the other end of the scale, very high precision will incur very high manufacturing costs. The manufacturing techniques will be specialized, hence the compatability will be low. Likewise, the performance will be susceptible to misuse and intolerant of anything but its prescribed operating environment, hence it could be low. Somewhere between these extremes is the precision level that satisfies the performance and cost parameters desired.

profile. The second level refers to the inspection tasks that are conducted in order to ensure that a particular assembly task has been performed—it will not necessarily check the functional result of the task, just the fact that it has or has not been performed. The third level is that of functional inspection, which ensure that after each assembly stage, what has been performed has met the quality conditions of the process and that (as far as can be determined) the active components function according to specification. An example of this is the dispensing of hot melt sealant near electronic components. The second-level inspection ensures that the components are there and that the hot melt sealant has been deposited. The third-level inspection task is to check that the electronic components have not been damaged by the heat of the hot melt sealant. Proper consideration at the design stage will provide the various subassemblies as well as the product with easily accessible electronic or visual inspection points for assessing the second- and third-level qualities of the units.

One major criticism of many products that have been designed for automated assembly is that they are still designed from the human point of view in terms of both procedures and work ethics. If FASs are to succeed and if the computer automated factories (CAFs) of tomorrow are to evolve, then it is necessary that the techniques and methodologies used *must* be compatible with machine functions and capabilities. Machines often have motions that are impossible for a person to duplicate—for instance, the rotary wrist of a robot can generate a controlled centrifugal force on a component held in its gripper. A human could not replicate this action in any prolonged and/or precise manner. As computer-aided design (CAD) becomes more and more widely adopted and sophisticated, the stage will be reached where machines design products for machines—which should resolve the present conflict of work ethics.

The final aspect of product design that could effect the assembly process is that of whether the product is to be repairable or disposable. If the latter, then the joining processes can be of a permanent nature which will invariably result in a lower cost product than if it had been designed and assembled with repair and refurbishment in mind.

People

The assembly process has traditionally been labor intensive, which with the human propensity toward variability of output, inconsistency of performance, and fatigue, has more often than not resulted in chaos and inefficiency. The corollary is that manually assembled products require and support a lot of nonproductive activities (such as inspection and rework) in order to cover-up for the inherent failure of humans as productive elements. Humans can also be very malevolent (if they lack job satisfaction) and by acts of deliberate sabotage or pranks—to break the monotony—can at best cause the rejections of assemblies or at worst the total shut down of the FAS.

The majority of tasks within an FAS can be separated into two groups, those that should be done by machines and those that should be done by humans. Allocation into the first category is determinate, whereas the second category allocations are nearly always subjective.

The first category tasks are those chosen by the need for predictability, consistency, and reliability of the process and product. Hence the use of automation (in its widest sense) for stock control and the conversion sector. Humans are used when management has decided that the inherent risk of using them is outweighed by the financial and technical constraints of an automated solution.

The assembly process requires the attributes of judgement, dexterity, flexibility, and low cost. Humans that are employed in assembly work do not always supply the right qualities, since their judgement is fallible, their dexterity limited, their flexibility is not uniform, their total cost is often more than its face value, and if strength is required, then that of the human quickly fades. Therefore, it can be reiterated that assembly by humans is not compatible with a deterministic environment, since their susceptibility to emotion and fatigue results in performance, output, and quality being variable, inconsistent and not predictable in absolute terms.

Why then should humans be used within a FAS? The major answer is that, external to the tasks directly associated with the conversion of raw materials and purchased components into value added products, there are many tasks that cannot (yet) be performed satisfactorily by automation. While a lot of these tasks require machines to generate and manipulate information or to do the "number crunching," it is human intellect that makes all of the decisions. Such tasks within the FAS are those of management and policymaking which endeavor to ensure that the FAS functions competitively. The difference between this and productivity is that the former means selling the goods that are assembled, whereas the latter relates only to the conversion process.

Other categories of human oriented tasks are those where the response to a given situation demands the variability of human capability, i.e., maintenance; where automation of the task would drastically outweigh any financial benefit, i.e., cleaners; or where the human attributes of speed, delicacy, and low cost are desirable, i.e., in the use of humans to assemble electronic consumer goods. In summary, the human is best occupied in tasks where intellect, variability, delicacy, or low cost are the premier requirements.

Two of the overriding priorities of BRSL's brief[2] were that automation should not be used for automation's sake and that the working conditions and job opportunities within the FAS were to be paramount. An example of adherence to these requirements is the design of the work stations, which are biased toward being comfortable and compatible with human ideals. The layout and environment of the work stations is psychologically pleasing and the natural boundaries created by the conveyors optimize the social groupings. The net result is that the humans are situated in an environment where working conditions are excellent and the opportunities for job satisfaction are given implicitly and explicitly. The former is offered by the fact of working in a dynamic, new, technology-based factory, and the latter by the sense of competition that will be stimulated between the various work areas.

The human tasks, within the FAS evaluated by BRSL, are best explained by starting with the work centers, of which there are two basically identical units. Each work center has 38 work stations grouped in six work areas. There is one operator for each of the 20 manual work stations per work center, and an additional operator, whose task is that of trouble shooting the automated PCB line. Since each work center will work a double shift, the total number of work station personnel will be 84.

Each work area will have a supervisor—so that interaction management can function throughout the FAS—who will have ultimate control of that work area's schedule. The computer-generated schedule will be discussed regularly by the work area supervisors and sales personnel. The work load will then be distributed (by whatever "barter system" evolves) between the work areas. Within their own domain each work area supervisor can play with the schedule through the local terminal, and while the work cannot be removed from the schedule (until complete) it can be removed from the work allocation of one work area to that of another, provided that there is mutual agreement. With the 12 work areas and double shifting, the total number of work area supervisors will be 24.

The start-up and learning curve eventualities of the FAS will require far more direct labor personnel at the beginning than when it is operating at the steady state condition. A factor of 50% was therefore applied to the "direct labor" force, giving a total of 164 persons.

The viability of the FAS depends upon all of the systems functioning to specification whenever they are needed. Machines and system failures will and do occur. The automated equipment and control systems will be designed as modular units, so that a small team of trained personnel can effect immediate recovery for the majority of the FAS failures. For this task two persons per center shift are considered adequate.

In addition to the eight emergency service personnel, a highly qualified team will be provided to repair the replacement modules, to organize and expedite preventive maintenance, as well as to carry out development and commissioning functions. Since this team will not be directly related to the production schedule, five persons on the major shift were deemed sufficient. When the FAS is at steady state operation and information on local servicing and support facilities by the vendors of the capital equipment is known and proven, the number of team members can be changed.

Continuous evaluation and upgrading of the FAS's operating systems will be done by six industrial engineers, whose expertise will cover all of the disciplines used within the FAS. This team will work a single shift, unlike the three-shift team that will repair and maintain the building fabric, conveyor system and elevators, etc. These 24-hr service personnel will work alongside the "high technology" team mentioned earlier and will have to be extremely competent and capable, as they will have to cope with new installations, sudden and unpredictable occurrences, and factory alterations. This team, which will include both mechanical and electrical specialists (and/or persons with multidisciplinary competence), will also be responsible for the preventive maintenance routines that are performed during the third *dead* shift, so as to enable the system's up time to be maximized. While the automated storage system, the elevators, and the conveyor system can all be serviced during this dead shift, work station maintenance will have to be conducted during working hours because of the inherent flexibility of the system. While it is considered sufficient that two persons per shift will be able to satisfy the demands of this function, it is acknowledged that members of the various teams could be combined to get the FAS operational if a major failure occurred.

The stores related functions of goods inward and goods outward require fairly large teams of 14 persons per shift covering the two-shift operation of the FAS. Quality assurance, on the other hand, will operate a single-shift service with 12 personnel. In addition to the defined functions, there will be a group of some 36 persons working primarily a single shift that will fulfill the FAS's needs for secretarial, reception, canteen, and cleaning staff.

The overall responsibility for the operation of the entire FAS (independant of shift) must belong to a single person—the director—who ideally will be a Ph.D., P.Eng., with an extremely developed and fine-tuned business acumen. Reporting to this person will be two managers, one for each operational shift, who will run the FAS on a day-to-day basis through the computers and the work area supervisors. The staffing levels discussed are summarized in Table 15.

6.1. People and Robots

The major technology that distinguishes the FAS from the ordinary manufacturing plant is that of microelectronics, with the most controversial symbol of this new mega industry being the industrial robot. The introduction of robotics into industry

Table 15
Staffing Levels for the FAS Evaluated by BRSL

Personnel function	Shift 1	Shift 2	Shift 3
Work station operator	40	40	00
PCB operator	2	2	
Work area supervisors	12	12	
SUBTOTAL (direct labor)	54	54	0
With 50% increase number of direct labor	81	81	
Emergency service personnel	4	4	
High-technology maintenance	5		
Low-technology maintenance	2	2	2
Industrial engineers	6		
Goods inward/outward	14	14	
Quality assurance	12		
Shift manager	1	1	
Director	1		
TOTALS	126	102	2

Total factory staff	230
Undefined staff	36
GRAND TOTAL	266

in general and FASs in particular has resulted in a three-level psychological interface between humans and robots that has to be acknowledged and accomodated if the FAS is to make efficient use of this technology.

The first-level interface is the realization that robotics allow the relief of humans from the necessity of performing operations in extreme working conditions. While the removal of humans from alien working environments is undoubtedly a good social act, many unions see disadvantages to themselves from this trend of humanization of work. If a human job function is taken over by a robot—for whatever reason—then it could be that the human is redeployed elsewhere in the plant, performing a task that is outside the influence of that union. Since the robot has no allegiance to any (human) union, the number of union members in any one plant could decline. The situation is presently relatively calm, since these economically hard times have caused labor to grant many concessions to industry, amongst them the acceptance of robots working alongside or in place of union members. Nonunion plants have even less trouble, since while the workers may grumble, they are thankful to have a job and they do not have a say in the matter anyway. Generally the alternative to being reasonable (on both sides) and receptive to the adoption of robotics and other forms of automation is a plant shut down, which would definitely result in a jobs loss.

The second-level interface is the recognition by engineers and management that increased productivity can be obtained, provided that the work place is designed around the robot. In the late 1960s, Pehr G. Gyllenhammer,[14] the president of

Volvo, said that " . . . factory work must be adapted to people and not people to machines . . . " In order to maximize the productivity of the FAS, not only is that quote true, but also the paraphrase that "the workplace must be adapted to the robot and not the robot to the workplace." Adoption of this policy means that modification of a given component can result in a simple rather than a complex gripper being required to perform a given task.

It also means a rethinking, in machine terms, of what is required from a particular workplace by the appropriate engineer or manager. This is because often the method of doing a particular job is colored in the decisionmaker's mind by the way that (s)he would do it. What should happen is that the decisionmaker should think in terms of how the robot would best like to do it. For example, using the rotary wrist to spin off the excess slurry from an investment casting tree.

While robots are usually cited for their increased productivity, management should be aware of their use in coping with fluctuations in demand for the FAS's products. By using the robot's productivity capability profile, their productivity can be adjusted to match the market demand profile for the goods. This means that the FAS can ride out the fluctuations of the market without altering manning levels and so eliminating a major cause of friction between management and the workforce. It also saves a lot of money by not having to hire persons, train them, fire them in a down phase, and then have to hire and train others as the market trend varies.

6.1.1. People Problems

The third-level interface is the acknowledgement of the psycho-sociological problems that may be encountered by the human supervisor of robots. The adoption of robots has meant a change in job status for a number of humans, in that their function has changed from doer to teacher (robot supervisor and/or instructor). This change has caused a number of psycho-sociological benefits and problems to arise. The benefits are primarily increased job satisfaction, which manifests itself in the form of low absenteeism—where before the status change it was 40%, after the change it is zero.[15]

The problems occur because while the human operator/programmer is the most crucial element in determining the effective use of the robot within the work station, they are the least understood. Why, it may be asked, is there the need for any human involvement, since is not the purpose of developing and introducing automation so as to get rid of the slothful, costly, unreliable human from the shop floor. In contrast to the automated production of commodities that are not subject to change in demand, and where large amounts of capital equipment are occupied by very few human workers/supervisors, the FAS represents a veritable maelstrom. This is due to the continuous change in demand for the FAS's products and as a consequence (in one form or another), the processes are in a permanent state of flux. Hence, there is the need for frequent programming and debugging of the system by humans for any number of reasons.

The psycho-sociological problems occur at two distinct sublevels. The first is the short-term, functional, man/machine system level where the human is in close proximity to the robot, and functioning in any of the four modes of supervision/interaction. The first of these modes is the planning of the robot program. Here the supervisor must think in robot terms—what program and what maneuvers will give the highest productivity and best quality. The second mode is that of the teaching of the robot, where the pattern has to be entered, tested, debugged, and proven before it can be run under operational conditions. The robot can seem to be "awkward" or to have a "mind of its own" during these preoperational phases. The third mode is that of monitoring the robot as it performs the tasks. Here the supervisor can be doing two jobs at once, since apart from the monitoring of the robot's performance, the human is often required to perform another separate task. If automatic monitoring equipment is installed, then the supervisor may wonder which of the two workers is *really* being monitored—the robot doing the task or the supervisor on the basis of how well the task is being done. The fourth mode occurs when the supervisor intervenes in the normal operation of the robot and enters the direct control loop of the robot for emergency shut down, maintenance, or repair. Here, in very big systems, the identities of the robot and the supervisor can become confused. Fortunately, the robots that tend to be used within FASs do not have this effect.

The second sublevel is that of the long-term psycho-sociological interaction, where the supervision assumes a different cast. Again, there are several modes of occurrence, with the first being the possibility that the role of robot supervisor is alien to the particular human. Alienation of the human worker/supervisor/programmer can originate through several factors. For instance, because the workplace is liable to be hazardous and since the electronic displays will be at a remote place from the scene of action, the supervisor is most likely to be spatially remote. Also, as the supervisor's instructions to the robot will not be instantaneously obeyed, the human will experience a feeling of temporal remoteness. A classic example of these states of alienation are given in a study of a fully automated steel plant in the Netherlands.[16]

At the Hoogovens steel plant of Royal Dutch Steel, a study was conducted to determine the cause of a productivity problem on an automated strip rolling mill system. It was determined that the humans who were monitoring the system were unable to tell if the system was failing and exactly when to intervene to take over control from the computer system. So unsure of themselves were they, that they sometimes left the control stations unmanned. Also, as they were removed from the environment of the mill floor (spatially remote) and since the operations were enclosed for safety and environmental reasons, the supervisors lost all visual contact with the system, which disoriented them still further. This meant that they were unable to assess when the computer system was failing to effectively control the operation through its complex of computer programs and electrical input sensors. Hence, the supervisors were temporally remote as they were unable to measure directly the results of any changes to the program. The personnel developed the attitude of standing well back and intervening only when things were clearly going awry. This was unfortunately usually too late as productivity had already dropped to levels even below that of plants with conventional control. So automation had led to lower productivity and operator alienation simultaneously.

The cures were as complex as the disease, with the control methods (including the way that the programs operated in process terms) having to be made explicitly clear to the supervisors. Further, all visual clues had to be made available to them so as to guide them in understanding when things were not working properly. The training, of course, had to be more extensive and the theory of the control system had to be carefully taught, even though, because of the complexity of the process physics, this was not easy nor really practical.

If a human does not practice a skill, it degenerates; hence, if the supervisor spends most of the time overseeing the robot, it follows that when called upon to perform that task personally, it is unlikely that is will be performed well. Thus, that person's neuromuscular skill(s) is not only usurped by the robot, it also deteriorates, which can lead to resentment by the human.

It should be noted that this decay of skills due to nonuse does not occur with the robot, since once it has been programmed the information is in some form of battery-backed memory. So, even if the main power is switched off for several weeks the information is retained. The use of EPROMs (eraseable and programmable read only memory) means that the instructions will always be present and the skill level never varying except during the optimization period of the task performance *or* if the EPROMs are erased.

The last of these alienation factors occurs because the robot is faster, more accurate, and more reliable than the human. Further, since robots do not "share their feelings," it is an easy human failing to attribute qualities to them that they do not have—wisdom, judgement, and worthiness. Hence, the robot's personality is both seductive and alienating.

The second long-term cast change occurs because the human supervisor is bound to perceive the striking difference between productivity in machine terms and productivity in human terms. In machine terms, efficiency is the criterion whereby the most product is obtained for the least raw material and/or effort. In machine terms, more order and less variety makes for greater efficiency, i.e., least effort is best, whereas in human terms variety is good and even "healthy." The shift in emphasis from physical and mental work to primarily mental work means that the new generation of robot supervisors are apt to exercise their bodies only after work. This results in their neuromuscular and occupational work probably becoming completely unrelated.

6.1.2. Robot Personalities

The idea that robots have their own personalities is no longer scoffed at, since there is now evidence to prove that two identical robots will not perform identical functions when using the same tape. In the same vein, it has been semiseriously suggested that a practical way of evaluating the performance of a given robot against potential applications is for each robot to have its own *resume,* so that any potential "employer" can evaluate the applicant for the job.[17] The trend toward the personalization of automation in general and robots in particular is shown by the predominantly Japanese tendency to paint and name the machines so that they can be incorporated as dumb members of the social group of the work area.

6.2. Unemployment, Redeployment, and Employment

It is often claimed (in the popular press) that automation and microelectronics will cause massive unemployment within the manufacturing industries of the countries that adopt the technology. In fact the reverse is true in that the countries and companies that *do not* adopt the mega industry of microelectronics in all of its guises will fail to be competitive and so fall into stagnation, with disastrous results for the population.

Another truth is that no matter how automated a plant or industry becomes there will always be some human factor involved, albeit that it will often be only for very intelligent and/or competent humans. Historically, it has been proven that machines never replace humans within the manufacturing industry, they simply remove the need for humans to perform certain tasks; also, it has been proven that as new technologies are created they generate new problems that are in turn solved by humans, and often the solutions involve humans to achieve their targets. It was once considered that the steam engine would cause great unemployment for men—and so it did for man, the laborer—but it created new challenges and new opportunities that would never have been if it were not for the first mega industry of steam power. Early computers were hailed as automatons capable of replacing humans in many areas of accounting, calculations, record keeping, and even decisionmaking. While they eliminated a lot of number crunching chores, the computer's limitations were soon realized. At the same time, computers spawned whole new industries and opportunities in engineering, programming, computer operations, computer maintenance, sales, marketing, and so on. Many of the problems stem from the transition period that most of the industrialized nations are now entering as they move into the status of being postindustralized nations. This phase is exemplified by population moves from one industry to another. It is similar to the migration from the agricultural industry to that of the new manufacturing industry that occurred during the first industrial revolution. Nowadays the transition is from the manufacturing industry to that of the service or information industry.

The adoption of microelectronics-based FASs and their interaction with society will manifest itself in a three-pronged short-term reshuffling of humans within the manufacturing industry. The three drivers that will cause this rearrangement are unemployment, redeployment, and employment. The unemployed will be those humans who cannot or will not accept the changes brought about by the force behind the FAS—micro-electronics. Other unemployment will be caused by companies who have not adopted the new technology for their products or processes, with the result that they will fail to be competitive and so go bankrupt. The number of companies that adopt microelectronics purely as a means of reducing their labor force will be very few and far between, since one of the results of adoption is improved productivity. This means that the company's products are more competitive, with the net result that in order to satisfy the increased demand for the products, the company will have to employ more staff, albeit in the non-production-related activities.

Humans must learn to work with and to trust microelectronics and not to work against them if automation (in its widest sense) is to result in tangible benefits for mankind. Many humans see robots as threats to their jobs, which while true in many cases enables good management to use this opportunity to provide training for these displaced workers in the new technologies and transferring or promoting them from within to higher status and higher paying positions. This was mentioned earlier with the transfer of the human from the boring, degrading job in an often hostile environment to the higher status position of robot supervisor.

Redeployment often means retraining of the human such that the new task can be performed confidently and efficiently. Retraining should be considered as a positive gift of microelectronics and not be regarded as demeaning. In fact, retraining at intervals of ten years will come to be considered the norm for all employees within the manufacturing industry, no matter what their station. The reason is that any information gained during their previous educational sessions will be considered obsolete within a decade. The educational process will most likely take the form of continuous upgrading of information, linked with structured sabbaticals at appropriate time intervals. While this philosophy will allow a person to transfer from one profession to another—providing that the person has the intellect—it means that eventually there will be no such thing as an apprenticeship system for equipping a person with knowledge suitable for a lifetime's employment.

One serious question that is yet unanswered is that of de-skilling. While it is relatively easy to accept that individuals will have to undergo a transition from the skills of experience to the skills of analysis, as well as undergoing regular retraining, it is more difficult to come to terms with the total loss of those skills through the lack of practice or through obsolescence. One trade union based fear is that skills will become polarized with the industrialized developed nations monopolizing the high-skilled occupations, while the developing nations are contracted to perform the lower-skilled work.

The development of FASs and the wholesale adoption of microelectronics *will* mean employment, although it will not necessarily be directly related to either the industry or the technology. Employment will come about through new industries that are created by and through the technology. There will be an increased need for persons in the service and information industries that identify the post-industrialized nations.

The adoption of microelectronics simultaneously permits removal of humans from industries that cannot be competitive if they continue to use humans and encourages the development of industries that are human biased. There will also be a proliferation of the so-called "electronic cottages," in which individuals will be connected to computers and/or other cottages through a wide based and sophisticated communications network, as and when they need to gain or transfer information. This return to "home" industries will greatly improve the lifestyle of those so employed, as they will invariably be doing what they are good at. An economy that is supported by microelectronics based FASs and CAFs will allow and encourage this working environment.

Robots

The predominant genus of automation associated with the FAS is the robot. These machines are the most controversial symbol of the mega industry of microelectronics, with additional confusion caused by the claims and counterclaims of rival robot manufacturers as they set the robot world ablaze with truths, misinformation, and conjecture about their products and industry.

Within the manufacturing industry, robots of all shapes and sizes are replacing man the laborer in the same way as computers are taking the number crunching burden from man the mathematician. They are used to perform functions that are injurious to the health of humans, e.g., spray painting, adhesive application, arc and spot welding, and fettling in foundries. They are also replacing humans in tasks that are boring and fatiguing as well as being used within environments that are definitely not conducive to humans (e.g., areas of high radioactivity or those zones requiring supreme cleanliness). Finally, they are being used more and more for assembly related tasks, as the technological level of the robots increases and as the true benefits of robotics is realized by industry.

Robots mean different things to different people and nations. They are sold under many different guises and in order to bring some control to the industry, the Robot Institute of America (RIA) formulated the following definition:

> A *robot* is a *reprogrammable* multifunctional manipulator designed to move materials, parts, tools or specialized devices through *variable programmed* motions for the performance *of a variety of tasks*.

This definition, or a slight variant of it, has been adopted in the majority of countries that manufacture and/or use robots. The Japanese have some six definitions of robot, four of which are covered by the RIA definition.

Within the definition, robots come in two distinct categories—nonservo and servo. The former describes a robot whose movements are set manually through adjustable stops on each and every axis. Consequently, the number of possible movements (per program) is twice the number of axes. Servoed robots are those whose movements are totally controlled by computer and hence there are many thousands of theoretical movements possible within one program. This is because a movement is any distance between two points within the extremities of each and every axis. Additionally, there are no end stops to be adjusted and program changes both large and small are generally easily performed. Both nonservo and servo robots are invariably microprocessor controlled, although in the former case the control only relates to the sequencing of the actions. In the latter case, every action, function, and task is initiated and supervised by the microprocessor.

There are two types of motion control. The first is point-to-point, whereby the two extreme points of any action is keyed into the computer and the motion between these two points occupies a random path. This randomness, however, can be contained by the use of certain firmware strategies (in some robots), which specify the sequence and action of each axis move in order to obtain the desired action. The alternative mode of motion control is that of continuous path, whereby every point along the pattern is recorded and can be reproduced without fail, within the tolerance band of the robot. In both forms of control, motion is usually in three-dimensional space and can involve the interaction of up to seven or eight degrees of freedom, this being the total sum of the robots axes of movement.

Robots are manufactured in a number of configurations and are classified by the "gross" motions that are possible by their major axes. While the terminology of robots is beginning to be defined and accepted, there are at present no agreed upon industrial standards for assessing a particular robot's performance. Hence, the objective comparison of robots from a number of manufacturers is not yet possible to any standard that would be acceptable to the industry.

In general, there are four configurations of robots: (1) Cartesian, (ii) polar, (iii) cylindrical, and (iv) arm and elbow. These are shown symbolically in Figure 28. An in-depth study by Ioannou and Rathmill,[18] showed that 50% of all robots available, at the time of their investigation, were of the arm and elbow style. As is shown in Figure 29, the next most popular types are those of the cylindrical and turret forms, followed by the Cartesian mode, which only occupies 6% of the total installations investigated.

An analysis of the American market shows that Unimation, Cincinnati, DeVilbiss, and ASEA have some 67% of the market share between them. Unimation produces Unimates and Pumas in a ratio of about 2:1. (turret:arm and elbow). Therefore, the comparative figures for arm and elbow are about 45%. The combined value for both turret and arm and elbow are[18] about 64% in the U.K., which compares well with the above mentioned 67%.

The popularity of the arm and elbow configuration can be understood for tasks that require a lot of manipulation, such as in machining or painting. However, for assembly, where the majority of the tasks are pick-and-place motions with precise positioning in the horizontal and vertical planes, the Cartesian configuration would seem to be more appropriate. The comparative arguments are those of the robot-centered working envelope of the arm and elbow configuration versus the external working envelope of the Cartesian device, and the cost and technology needed to control the many segments of the complex option in order to give a vertical or horizontal motion of the gripper versus the simple direct action of the more rigid system.

The drive systems of robots have traditionally been hydraulics for servoed sys-

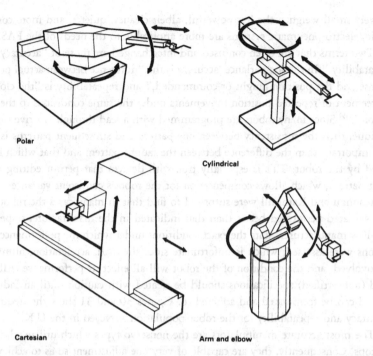

Polar

Cylindrical

Cartesian

Arm and elbow

Figure 28. Basic configurations of industrial robots. (Reprinted with kind permission of Production Engineering Research Association.)

tems and pneumatics for nonservoed systems. The trend is now toward electric and electro-pneumatic systems as the main power drives, especially with the newer models of robots which tend to be smaller, though more sophisticated, than their earlier and more powerful predecessors. Hydraulics with their inherent problem of leakage are and will be used for robots that require a lot of power, i.e., the big Unimates and Cincinnati's. However, since assembly tasks generally require the manipulation of only

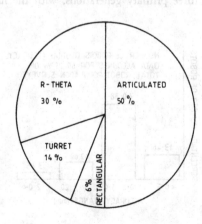

R - THETA

30 %

ARTICULATED

50 %

TURRET

14 %

RECTANGULAR

6 %

Figure 29. Classification of known commercially available robots by configuration. For comparison with Figure 28 and text, read polar for turret, cylindrical for R-theta, Cartesian for rectangular, and arm and elbow for articulated. (Courtesy of Ioannou, Rathmill, and IFS.)

relatively small weights, the less powerful, albeit cleaner, quieter, and more compact electric/electro-pneumatic systems are more applicable to the needs of the FAS.

Two terms that are often confused and interchanged are those of "accuracy" and "repeatability." In robotic parlance, accuracy is the "difference between actual position response and the position taught or commanded," and repeatability is "the closeness of agreement of repeated position movements under the same conditions to the same location".[19] Since many robots are programmed with a lead through or drive through technique, then the consistency between one pattern and subsequent patterns is often more important than the difference between the taught pattern and that which is performed by the robot. This is especially true with devices that permit editing of the taught pattern, which allows compensation for the robot's consistent variance.

Ioannou and Rathmill were surprised to find that in many cases the robots that they evaluated performed better than that indicated in the manufacturer's specifications. Few manufacturers state the exact conditions under which the performance specifications were assessed, which is unfortunate since the load, acceleration, number of axes involved, and the condition of the robot will all effect its performance. All publicized (noncertified) specifications should be treated with caution until an industrial standard can be formulated and adhered to. Figures 30 and 31 show the frequencies of accuracy and repeatability for the robot population surveyed in the U.K.

The most accurate manipulators are the nonservo types which utilize adjustable end stops. Consequently, they are capable of very fine adjustment so as to achieve the desired performance. The choice between a nonservo manipulator and a fully servoed robot for a given task depends upon the price, convenience, and flexibility demanded by the assembly system. It is often the case that the best selection procedure is to specify what is required and to then try and get the best fit, initially from the manufacturer's specifications and then finally after a demonstration of the selected units performing tasks identical or similar to those anticipated for the application. A check list as shown in Table 16 is a good starting point in the selection procedure.

Robots are classified into technological "generations" (like computers). There are three primary generations, with the first relating to simple pick-and-place units—

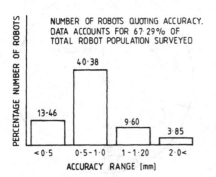

Figure 30. Frequencies of industrial robot accuracy characteristics. (Courtesy Ioannou, Rathmill, and IFS.)

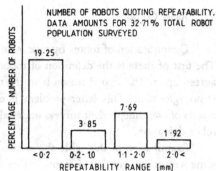

NUMBER OF ROBOTS QUOTING REPEATABILITY.
DATA AMOUNTS FOR 32·71% TOTAL ROBOT
POPULATION SURVEYED

Figure 31. Frequencies of industrial robot repeatability characteristics. (Courtesy Ioannou, Rathmill, and IFS.)

which do not always fit in the definition mentioned earlier—through the second generation which have "hand and eye" coordination, to the third which has the ability to "think," i.e., make decisions, plan and execute tasks, and interpret information received via sensors and other means. The general concensus is that the present generation of robots is somewhere between generations 1.5 and 2.5. Unfortunately this classification system is not precise enough since there already exist a few robots whose microcomputers and software permit them to perform decision-based tasks. The "intelligence" of the robot is, however, a function of its hardware, software, and the skills of the programmer. There are at present no known "free will" computers or robots at large in the manufacturing industry.

Table 16
Check List for Task vs Robot Compatibility

Positional accuracy
Maximum load (including end effector)
Velocity (do not confuse angular and linear velocities)
Configuration
Strokes and numbers of servoed axes
Strokes and numbers of nonservoed axes
Drive system
Motion control
Is system microprocessor controlled?
Number of inputs
Number of outputs
Degree of intelligence (very subjective, and is really a function of the I/O's)
Teach modes—pendant/drive through/lead through/coordinate entry
Memory size (number of pattern points and extensions)
Number of patterns held
Can it be linked to a CNC system?
Are there self-diagnostics?
Floor size and weight
What are the maintenance and servicing facilities?

7.1. Robot Census

Quantification of robots by task is an extremely difficult exercise on two counts. The first of these is the definition of robot, which has not been globally accepted or agreed upon. The second reason is in the task definition itself, for which again there is no agreement. This latter problem is exemplified in Table 17 which shows the results of two independent surveys undertaken in the United States using the identical robot definition.[20]

Examination of this table shows that the numerical totals for the robots is the same within 2%, yet there is considerable difference between the surveys for each task group. In one case, for assembly, the difference is 1,000%. It follows, therefore, that the classification is based upon subjective interpretation of the tasks involved. Under these arbitary rules, does fettling come under foundry or machining?; at what degree of accuracy does palletizing/packing become assembly?; and is enamelling the same as painting or does it go under "other"?

The number of statistics on the population of robots is enormous and contradictory. The general concensus is that the growth will be exponential over the next five to ten years with the requirements for spot welding, arc welding, and surface treatment applications declining after 1985. The growth applications for robotics will then be those of assembly and inspection.

In the U.K., there are some 32 companies that manufacture and/or sell 91 types of robots. In the U.S., there are over 100 companies involved, whereas in Japan the number of robotic companies is about 140, albeit not all of their products confirm with the RIA definition.

The estimated world population of robots in the RIA definition was around 37,500 units at the end of 1983. Figure 32 shows the distribution by country as compiled by the British Robot Association under their definition, which closely matches that of the RIA. For a number of countries the claimed population varies dramatically dependent upon the source of information. For instance, in 1981 the

Table 17
Estimated Robot Population of the United States by Task in
1981

Task	Daiwa securities	RIA
Foundry	615	840
Spot welding	1,435	1,500
Loading	820	850
Assembly	410	40
Painting	615	640
Other	205	400

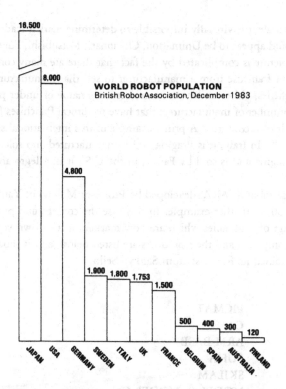

Figure 32. The world robot population as of December 1983. The census was conducted by the British Robot Association using their own definition of a robot—which closely matches that of the RIA. [Courtesy of IFS (Publications) Ltd, Bedford, England.]

estimates varied from 4,100 to 5,000 for the U.S., from 10,000 to 14,000 for Japan, and from 600 to 1,700 for Sweden.

The United States is the country with the longest experience of industrial robots. It is also the home of 25% of the world's robot manufacturers, including the traditional market leaders Unimation, Cincinnati, and Prab, who in 1980 with Trallfa/ DeVilbiss, ASEA, and Copperweld Robotics accounted for 95% of the total revenues of that market. By late 1983, this share will have fallen to under 60% (for those companies) for two very important reasons. The first is that there has been a large influx of companies into the market and secondly, the market has required, and these newer companies have tended to supply, lower priced, higher technology based systems. This is not to say that the traditional robot companies are not supplying the market needs, only that a lot of the newcomers are able to provide or respond quicker.

Thus the trend is toward the intelligent robot. These robots will exhibit the most substantial growth through the 1980s, with the market share for the manual manipulators and nonrobots declining significantly in value terms. By the end of the decade, sophisticated intelligent robots will constitute 45% of the total market value.

On a world scale it is virtually impossible to determine market leaders, although the "big six" would appear to be Unimation, Cincinnati, Mitsubishi, Fujitsu, Hitachi, and Seiko. The picture is complicated by the fact that there are many companies that have the license or franchise from a manufacturer to sell that manufacturer's robot(s) in particular countries, either under the manufacturer's name or under private label. There are also a number of manufacturers that have reciprocal franchises for each other's robots in their own countries. A prime example of this internationalism is give by the Pragma robot.[21] In Italy, it is Pragma and is manufactured and sold by DEA; in the U.K., it is Pragma and is sold by Fairey; in the U.S., it is Allegro and is sold by General Electric.

The Japanese robot SCARA, developed by Professor Makino of Yamanashi University in 1978–80, is another example. In this case the concept and prototype were funded by a group of companies which are now marketing their own versions of the SCARA. These companies and their products are listed below, and it should be noted that IBM has obtained its franchise from Sankyo Seiki.

Nitto Seiko	PICMAT
Yamaha	CAME
Hirata	ARM-BASE
Pental	PUHA
Sankyo Seiki	SKILAM
NEC	MODEL B, MODEL C
IBM (U.S.A.)	7535
Fujitsu	(to be announced)

What is evident is that there is now a large choice between robots from different manufacturers, which allows the purchasers to obtain a robot that more closely matches their specifications and price. The specification can be matched by a supplier having a range of robots to offer (not necessarily of the same manufacturer), but price is a function of the robot manufacturer's costs and desired profit levels. The prime cost of the robot to their manufacturer can often be reduced by setting up a manufacturing company in the target country or offering a license to an existing company within that country. In this way, robots can be reduced in price through economies of scale due to an increase in sales. Also, because there is now a *national supplier*, there is only the need to import the information to build the robots and not the hardware itself.

Table 18 lists all of the franchises and licensing arrangements known to the author, showing the scale and complexity of the relationships. It is important to realize that many of the relationships listed refer to the movement of "assembly robots," which have been developed by many of the newer companies. The arrows indicate the direction from exporter to importer, although in many cases the exchange is bilateral.

Table 18
*Franchises and Agreements between Companies Marketing
Robots as of March 1983*[a]

Japan		U.S.A.
Dainichi Kiko	→	GCA
Fujitsu Fanuc	↔	General Motors
Hitachi	→	Automatix
Hitachi	→	General Electric
Komatsu	→	Westinghouse
Mitsubishi Electric	→	Westinghouse
Nachi Fujikoshi	→	Advanced Robotics Corp
Murata Machinery	→	Prab Robotics Inc.
Sankyo Seiki	→	IBM
Kawasaki	↔	Unimation
Yaskawa Electric	→	Hobart Bros.
Yaskawa Electric	↔	Machine Intelligence Corp
Yaskawa Electric	→	Bendix
Tokyo Electron	←	Automatix

Italy		
Basfer	→	Nordson
DEA	→	General Electric
Olivetti	→	Westinghouse

Belgium		
Fabrique Nationale	←	Prab Robotics Inc.

Finland		
Nokia	←	Unimation

West Germany		
Nimak	→	United Technologies
Volkwagen	→	General Electric

France		
Renault	↔	Ransburg

Norway		
Trallfa	↔	DeVibliss

U.K.		
Remek Microelectronics	↔	Copperweld Robotics
Hall Automation	→	Otto Durr

U.S.A.		
Thermwood	→	Binks
Thermwood	→	Cyclomatic
Thermwood	→	Didde Graphic

Japan		U.K.
Yaskawa	→	Lincoln Electric
Hitachi (Process)	→	Lansing Robots
Dainichi Kiko	→	Dainichi Sykes

(Continued overleaf)

Table 18 *(Continued)*

Japan		U.K.
Star	→	Haynes and Fordham
Sankyo	→	IBM
Taiyo	→	Haynes and Fordham
Shin Meiwa	→	Dale
Fujitsu Fanuc	→	600 Group
Fuji Electric	→	Haynes and Fordham
Hitachi (painting)	→	Haden Drysys
Seiko	→	Airstead
Hirata	→	Airstead
Harmo Robots	→	Harmo Robots (UK)
Italy		
Basfer	→	Nordson
DEA	→	Fairey
Jobs	→	Fairey
Camel	→	Fairey
Gaiotto	→	Fairey
Elfin	→	Fairey
Basiach & Carro	→	FATA
Austria		
Limat	→	WIRA
West Germany		
Cloos	→	WIRA
Nimak	→	Telecamet
Norway		
Øglaend	→	Fairey
Sweden		
Electrolux	→	George Kuikka
Kaufeldt	→	Ringway Power System
U.S.A.		
American Robot Corp.	→	Rediffusion Simulation

*The marketplace is very dynamic; hence details in the table will change rapidly over time.

The reasons for the relationships are, for example: (i) joint venture involving products and ideas; (ii) sale for resale agreement; (iii) manufacturer to partner company for export to another country; (iv) elements of one product made by partner with relevant expertise for inclusion (sometimes) in joint products.

7.2. Robot Laws and Rules

As long ago as 1942. Isaac Asimov developed his Three Laws of Robotics, to define their relationship with humans:[22]

1. A robot may not injure a human or through inaction allow a human to come to harm

2. A robot must obey the orders given by humans, except where such orders would conflict with the First Law.
3. A robot must protect its own existence as long as such protection does not conflict with the First or Second Laws

While these laws form the basis of a code of conduct that defines and regulates the relationship of robots to humans, they do not define or regulate the introduction of one to the other. The introduction of robots into the workplace is more often resisted by management than by unions or the people that they represent.

The usual response by unions to new technology is one of willingness. The next most common response is one of opposition, but this is followed by adjustment by both sides such that acceptance follows. When a union does respond negatively, it is invariably because of the effect that the introduction of robots would have on its members. Once the union is convinced that its members will either not be adversely affected or that those who are affected will receive some offsetting benefit, the union opposition disappears.

Management is generally very conservative and prefers to use techniques and processes that are first of all proven, and secondly up to date. Also, if the adopted system does not work, then it is management that has taken the risk and will have to pay the penalty. Finally, management is reluctant to adopt systems that it does not understand, since if it is seen to be lacking understanding of or confidence in the new technology, then they are a soft target for criticism and abuse.

Clapp[23] has developed three laws that pertain to the installation and introduction of robots into the workplace:

1. Organizations may not install robots to the economic, social, or physical detriment of workers or management.
2. Organizations may not install robots through devious or closed strategies which reflect distrust or disregard for the workforce, for surely they will fulfill their own prophecy.
3. Organizations may only install robots for those tasks which, while currently performed by men, are tasks where the man is like a robot, not the robot like a man.

The first law means that amongst other things, the humans displaced by the robots would be guaranteed equally rated jobs, and that supervisors would not be penalized by down time while the robot was being debugged. It also means that the industrial engineer would not find himself out on a limb for recommending the robot.

The second law means that before any installation takes place all affected parties should be involved in discussions to determine and discuss the advantages and disadvantages that will either directly or indirectly result from the adoption (or nonadoption) of robots. There is also the hierachical problem to be resolved, since the normally accepted vertical hierachy with its structured problems and answers is suddenly uncertain, complex, and unyielding. This is because the long established measurements of efficiency and productivity are no longer appropriate. Also, the fact that the operator may be more knowledgeable than the manager further jolts the system.

The third law is again an obvious one, in that robots should be used to remove humans from tasks that are performed in noxious or hazardous conditions. However, removing people from one task and redeploying them at another can have a profound effect upon certain persons, especially when they are working in conjunction with a robot. This subject was examined in depth in Chapter 6.

Before a robot is installed, it must be chosen based upon its ability to perform against a prescribed set of requirements. However, there is a fundamental analysis that must be made prior to the scanning of numerous robot data sheets. The basic question that must be asked is: *Can a robot be used?* Implicit in the question is the fact that a robot is not the answer to every assembly situation. To answer the question, there are a number of factors that must be considered, and for each of these factors there is a rule. If the rule is applicable, then the use of robots for that application must be discounted. Tanner[24] identifies the factors and rules as follows:

Factor	Rule
Complexity of operation	Avoid extremes of complexity.
Degree of disorder	Disorder is deadly.
Production rate	Robots are generally no faster than humans.
Production volume	For very short runs, use humans. For very long runs use fixed automation.
Justification	If it does not make dollars, it does not make sense.
Long-term potential	If only one is needed, maybe you are better off with none.
Acceptance	If people do not want it, it won't make it.

The first factor has two extremes: (i) the case where the task is too simple to warrant the use of a robot or nonservo manipulator, i.e., a less complex solution (a cylinder and microswitch) would do, or (ii) the case where the requirement is for a lot of sensory perception and manipulation, i.e., a very complex situation.

The second factor relates to the attitude and orientation of the parts to be processed which must be in a predictable arrangement if the robot is to achieve a high success rate. The problems and solutions to disordered component presentation were discussed in Chapter 4.

The third factor is an expression of the payoff of sophistication versus speed of operation. A nonservo manipulator will operate at a high speed when linked with other handling equipment. A servo robot is less speedy than a simple manipulator because of the need to stop without the help of dead stops that give the simpler unit its speed. In general, it is found that robots will not operate as fast as a human per action, *but* the processing rate of acceptable assemblies will be higher per shift. Therefore, Tanner's rule is a reminder/check and not an exclusion rule.

The fourth rule is argued in Chapters 4 and 10 where depending upon the quantity of a given product to be processed per unit time, one of three techniques makes economic sense. For very small batches, the time required to change tooling and reschedule component deliveries from the stores may mean that it is easier and cheaper to use manual assembly stations. At the other end of the scale, the use of hard automation is financially and technically the better solution—better than the use of (say) robotics.

The fifth rule refers to making sure that the *total* benefits and problems that adoption of a robot will yield net out to a positive value. It is no good following the crowd or being pressured into robotics for the sake of robotics if it does not make economic sense. The need to investigate all of the benefits and environmental factors so that the complete picture can be drawn is the overall theme of this book.

The sixth rule relates to the purchase of a robot for a single, specific purpose that may be short lived. If management does not make the effort, then once the principle task has been completed, the robot gets forgotten—it gets written off by the accountants—and all of the associated skills gathered by the company for the robot become redundant and forgotten through nonuse.

The seventh rule has to do with communication. The direct labor force are worried about a jobs loss; management is concerned with production rates and quality; maintenance is concerned about its competence with the new technology; and the directors are concerned with profit and loss. If all of these groups do not communicate with each other, then any one of them could say that "robots are OK, but not here." If communication does take place, then there is the probability that the robot will be given a fair chance and not rejected out of hand.

7.3. Grippers

A gripper is a device that is often called the end effector of the robot. At the present time there is no industrial standard for the fixing holes of the end effectors, so consequently every gripper must have a different set of mounting holes to suit different robots. It is usually assumed that a gripper has typically no independent degrees of freedom. The reason is that it is anticipated that any degrees of freedom required will be provided by the other robotic elements, such as the wrist and/or arm. It has been recognized, however, that the use of certain devices (such as screwdrivers) will, because of their inherent degrees of freedom, result in the possible duplication of certain robotic capabilities.

The size and function of a gripper is directly related to the dimensional size, material, and weight of the components to be transported. The first parameter that defines the gripper design is that of the static load, which defines the operational envelope thus:

1. The component weight in terms of the capacity of the gripper determines the gripper's ability to accomodate the component. This limit is self-evident and

reflects the technical capabilities that have been designed into the robot/ gripper.

2. The gripper's ability to restrain the component's weight by frictional and shear forces determines the weight that can be held without slippage under static conditions.

3. The attitude of the gripper jaws during a particular maneuver controls the weight that can be moved. If the jaws are always at right angles to the gravitational force, then only a light force is required for component retention. If the jaws are parallel to the gravitational force, then the components weight is withstood by frictional and/or shear forces, the frictional forces being those that act directly between the gripper jaws and the component surface, while the shear forces come into effect when the component rests on the "top" of the jaws, such that the shear strength of the component/jaws resists the weight of the component. If the jaws are in any other attitude, then both the frictional and shear forces apply, but each to a lesser degree.

Figure 33 shows the three gripper–component orientations described above and indicates how the various restraining forces act. In all of these options, it must be remembered that the limiting force that can be applied to the component is the one that will distort and/or cause damage. It is also necessary to ensure that the loading is uniform

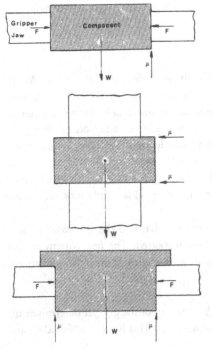

Figure 33. A component can be held in one of three ways by a gripper. The top illustration shows the component being held solely by the frictional force generated by the grippers. The middle illustration shows the condition where the component's weight is entirely supported by the shear strength of the gripper's lower jaw. The bottom illustration is a composite of the upper pair, in that the weight of the component is supported by both the frictional forces and the shear strength of the jaws.

across the "clamped" surface, otherwise a local highspot could cause localized damage to the component.

A gripper is used to move a component around in space. This generates dynamic forces on the component–gripper interface as the gripper accelerates and decelerates in the performance of its tasks. There are two sources of inertial forces that are generated:

(i) Centrifugal forces that act outward from the rotational center of the gripper arm as it moves through an arc. The centrifugal force increases in proportion to the radius and as the square of the angular velocity. It is resisted by the frictional and/or shear forces at the gripper–component interface.

(ii) Momentum is the force that is applied by the component and to the gripper as it is accelerated and decelerated. Again, the force is resisted by frictional and/or shear forces at the interface. The force is the multiple of the component weight and the acceleration (in g forces).

It should be remembered that inertial forces can lead to the "bursting" of materials and consequently, the design of *both* the component and gripper should be checked for this eventuality.

The third parameter that defines the operational envelope of the gripper is that of geometry, which refers to the maximum component dimensions that can be accommodated without affecting the desired performance of the gripper. There are four conditions that will cause instability of gripper performance because of component geometry:

1. where the component dimensions are such that the gripper cannot satisfactorily hold it;
2. where one or more of the component's dimensions are such that the center of gravity of the component causes a large tilting moment, either on the gripper or the component within the gripper;
3. where one or more of the component dimensions are such that the component cannot be maneuvered without collision; and
4. where the component is so small that the gripper cannot function properly.

The dimensional size and material are very influential on the mode of gripper action used. There are six broad classes of gripper 'mechanisms' used, albeit grippers can and do incorporate multiple mechanisms:

1. Mechanical clamping, is the commonest mechanism whereby pneumatic or hydraulically operated devices apply a surface pressure to a component. These grippers are available in three different styles:
   (a) Parallel jaws that hold a component between flat or vee'd surfaces. These devices can have one or two moving jaws.

    (b) Finger grippers (parrot beak) either wraparound the component or hold it at the very tip of the jaws.

    (c) Expansion/contraction grippers incorporate a flexible diaphragm, bladder, or other device, which when activated expands or contracts so that a frictional force is applied to the component. This mechanism is of particular use when dealing with delicate components or components whose geometric shape precludes the use of rigid clamping methods.

2. Magnetic clamping uses electromagnetism to hold the component. Obviously this system is only applicable to materials that can be held by a magnetic force and to components and work station environments that can withstand the magnetic field without incurring damage. One particular advantage of this system is that to a certain degree it is not component dependent.

3. Vacuum clamping is the application of negative pressure to components, so that they adhere to the gripper. The commonest style of vacuum gripper is the use of suction cups arranged in a pattern to suit the component with the vaccum being generated by an ejector or vacuum pump.

4. Piercing grippers puncture the component in order to lift it. This technique is used only where (slight) damage to the component is acceptable, i.e., clothing, etc.

5. Adhesive grippers are used for components that do not permit any of the above mentioned methodologies. They make use of a sticky tape to hold the component.

6. Universal grippers, because of the wide range of capabilities that are required of a gripper, are unlikely ever to be viable. What will and does occur is that a gripper is designed to accommodate a range of components within a family. This means that a set of pseudouniversal grippers could cater for all the needs of a work station.

The constructional material of the gripper itself is dependent upon the proposed duties of the gripper, the operating environment, and the components to be handled. Steel is the most obviously preferred material, since it is tough, easily obtainable, in a multitude of sections, and is compatible with the majority of manufacturing processes.

Aluminum is used where the gripper mass is of concern, or when a nonmagnetic yet tough material is required. Like steel, aluminum is readily available and can be processed easily. Plastics are used when handling a delicate component that could be marred or when it is necessary to ensure electrical isolation. The use of soft materials, such as rubber or foam, is restricted to expansion/contraction type grippers, or as padding on the contact surfaces of grippers manufactured from more robust materials.

The use of ceramics and other exotic materials is known for grippers that are used in hostile operational environments, such as may be encountered in a forging operation.

The technology that allows for the autochanging of grippers has been available for some time, and is not vastly different from that used on machining centers. The documentation of robots using autochanging is not numerous and tends to be confined to nonindustrial environments. However, as robots become more prevalent in assembly processes, where there will be the need for multigripper applications, the use of autochanging will become widespread.

One of the restrictions to autochanging is the interface between the gripper and the wrist/arm, which (unless umbilical cables are used) is used to transfer all of the operational and feedback circuitry between the two elements. Hence, self-sealing electrical and fluid couplings will have to be used so that the electrical, electronic, hydraulic, pneumatic, and optical systems that relay power and sensing capability to the gripper can easily be connected and disconnected.

BRSL[2] determined that the FAS would be handling six material groups each with their own gripper problems, as discussed below:

1. Plastic constituted the material for many of the components that required to be handled by machines. The problems anticipated are crushing and/or marking of the items.
2. Metal parts were the second most numerous group, but as many of the parts were fasteners, they required their own specialized tooling. Also because of the density of metal, very large components require very rigid and powerful grippers.
3. PCBs require grippers of great sensitivity to prevent damage.
4. Electrical components tend to be of very high density. In consequence, very powerful and rigid grippers are necessary.
5. Electronic components are the discrete items that are built into PCBs. The technology for efficient automatic assembly of these components already exists and it is considered unlikely that the components would be handled by means other than this specialized equipment.
6. Unclassified components are labels, foams, and so on, whose gripper problems (and solutions) cannot be itemized. Hence, they have to be resolved individually.

It was also determined that there were three families of geometric shape. There were round cylindrical components that can readily be held by means of a V-grooved gripper. The second family were prismatic and were usually capable of free standing on any face, and usually offered a number of gripping positions. The last family were the flat components, which have one dimension that is minute compared with the others.

The component weight range varied from less that 0.003 kg to approximately 1 kg. The distribution of component weight has already been shown in Figure 11, the complementary distribution of gripper "gap" sizes is shown in Figure 34 which indi-

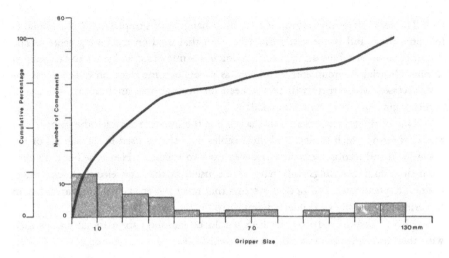

Figure 34. An analysis of components by BRSL showed that a gripper with a capacity of 40.00 mm could accommodate 70% of the items.

cates that while the range is from 0.00 mm to 130.00 mm, nearly 70% of the components handled could be accomodated by a gripper of 40.00 mm gap.

7.4. Economics of Robots and Grippers

Within the FAS, the price of a robotic system consists of three elements: the basic robot, the accessories, and the installation. As is shown in Figure 35, these costs are very dependent upon the application. The price ranges for the elements are:

1. Basic robot, between 40% and 70% of the total cost.
2. Accessories, between 20% and 35% of the total cost.
3. Installation, between 6% and 25% of the total cost.

In the past the reason for using robots were that they were useful in environments that were unpleasant for humans. While this continues to be true, it is not the primary reason for the adoption of robotics by the manufacturing industry in general. The major attraction is their flexibility and their capability for increasing quality and doing it consistently. Another justification, which is perhaps more far reaching and significant, is the make-up of the robots themselves. Since they are computer-based products—and are becoming more so every day—they both draw from and depend upon a data base. Thus, they are capable of generating management information, diagnostics, and tying into other forms of computerized automation.

This means that the correct robot in the properly designed (robot) workplace will

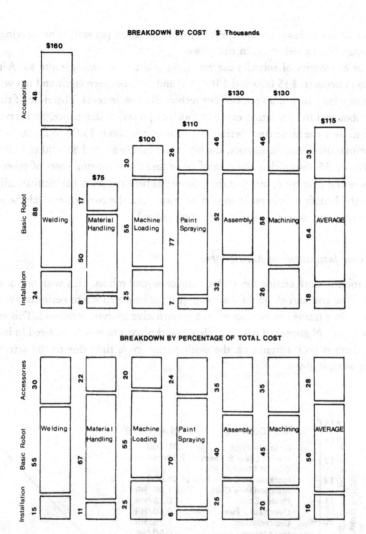

Figure 35. The total cost of any robot system is often ill defined and/or never computed. These listed values of typical robot systems by task give dollar and percentage values for the three system elements of basic unit, accessories, and installation. [After Mini-Micro Systems (April 1982) and EIU report 135 (December 1982).]

produce more output per working shift at an acceptable quality level than will a human doing the same task. It is interesting to note that the robot may not be any quicker than the human in terms of time per unit produced—in fact, it may be slower—but when all of the nonproductive time used by the human for physiological and social reasons is deducted from the available time, then it can be seen that while the human works faster, it is for a significantly shorter time and with an output that is usually variable in quality. Consequently, the use of a robot makes economic sense based upon the increased output of acceptable products per unit time, leaving alone any savings from a reduction in manpower.

The economics of robotics are attractive and are becoming more so. A typical robot costs between $35,000 and $100,000 and lasts between eight and ten years (on the assumption that it does not become technically obsolescent). Therefore, if the costs of the robot and the operating expenses, such as power, maintenance, and service, are amortized over the same time period, then the cost is about $5.00 to $6.00 per hour. In some manufacturing industries, the labor rate can be around $15.00 to $20.00 per hour. Figure 36 shows that the cost of labor and the operating costs of robots have both increased over time, but that the differential between them has dramatically widened to the benefit of robots. It should be noted that the curves can safely be extrapolated with time to give an up to date comparison.[25]

7.4.1. Cost Justification of Multigrippers

A robotic work station can include one or several robots, each with one or several grippers. The cost and choice of the robot–gripper combination to perform given tasks in a given time frame can be assessed in a quantitative analysis as follows. This analysis is of the use of N grippers for a generalized workplace which is considered in isolation so that there is no constraint on the work station cycle time due to the activities of adjacent work stations.

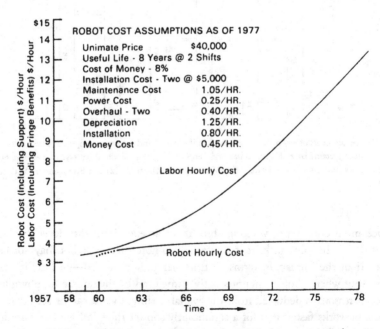

Figure 36. History of labor cost and of Unimate robot cost in the automotive industry.

For a given work station cycle time, it is self-evident that the total time taken in performing a number of actions is a function of the time taken for each action. Therefore, assuming steady state conditions:

$$t_c = \sum_{i=1}^{n} t_i$$

where t_c is the work station cycle time, t_i is the time to perform ith action, and n is the number of actions.

From the above, it can be seen that minimization of the nonproductive time within an action will result in a shorter cycle time. Minimization of nonproductive actions can be achieved through optimization of the work station functions and/or the use of more than one gripper. On the assumption that the optimization has been done, and therefore only the gripper aspects are to be investigated, the nonproductive actions are those which do not, for their whole duration, perform work on the component at the work station. These activities are:

(i) movement of robotic arm with nothing in the gripper.
(ii) movement of robotic arm while transferring subassemblies around the work station.
(iii) movement of robotic arm to and from the gripper storage station for gripper changeover.

From the above it can be seen that the time occupied by the nonproductive actions can be minimized by the use of:

(i) several robotic arms moving in concert.
(ii) careful design of the workplace.
(iii) multifunction grippers.

The use of multiarmed robots and/or multifunction grippers is only viable if the additional investment can be recouped through derived cost savings. Simplistically, we have:

$$I \leq P \sum_{i=n}^{n} (s_i \cdot c_i) - C_c - F - M$$

where C_c is the additional cost of control and programming per action, F is the additional direct and indirect cost factors per action, I is the additional investment at work station, M is the increased maintenance cost per action, P is the number of components produced in payback period, c_i is the cost per unit time for ith action, s_i is the time saving for ith unit, and n is the number of actions.

In addition to a quantitative rationalization of a generalized work station, there are a number of parameters that pertain to a particular work station. These must be investigated and resolved before the final decision regarding complexity of a robotic work station is made. These parameters are:

(i) the possibility of collision between individual arms of a multiarmed robot.
(ii) the working space in which the arm(s)/gripper(s) will be functioning.
(iii) the processing of more than one component through the work station—which may negate any cost benefit realized from using other than a simple robotic system

7.4.2. Economics of Gripper Design

Computerized automatic assembly may be regarded as the most advanced, complex, and sophisticated use of robots at the present time. In order to decrease the time occupied in assembly tasks to a minimum, it is necessary to give the greatest attention to the design of the grippers.

The design of the gripper has to be compatible with the geometric irregularities and complexity of the parts to be handled, as well as being optimally adapted to give it some flexibility for handling a range of components. Generally the aim is to use grippers whose inate flexibility enables their usage for gripping and manipulating *all* of the components of the assembly being processed. Unfortunately such a universal gripper does not exist today, not that it is not expected to exist in the foreseeable future.

The designs that come closest to satisfying the ideal of universality are the artifical hands (designed for robots, not humans) that have a variable and adjustable number of "fingers." One design exists at the Cranfield Institute of Technology in the U.K. A picture of the British Prime Minister Margaret Thatcher shaking hands with a Japanese robot was networked around the world in the fall of 1982. Unfortunately these devices are not really (at the moment) adaptable to the real world of industry and automated assembly processes.

Figure 37 lists the various gripper classifications that are available, the most common being the flexible style, which through built in mobility and adjustment accommodate a range of variable shaped and sized components. Another popular style of gripper is the multifunction unit in which a number—usually three—of different grippers are attached to a common base, so that the appropriate one can be activated and used when required. Generally each of the grippers is specially designed for a given purpose or range of parts and the use of the multifunction device allows the processing of a complex product without recourse to multiple work stations and/or changing single fixed grippers from a tool store as and when they are required. The processing time is therefore the sum of the task times plus the time required to rotate the multifunctional gripper as needed to present the correct gripper to the work.

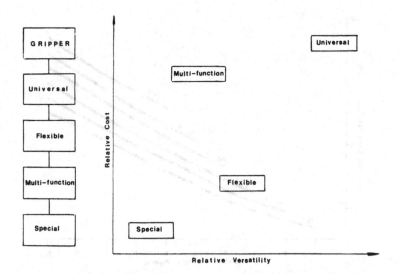

Figure 37. There are four gripper classifications. The value and applicability of each to the FAS is a function of their cost and versatility. The universal gripper has the highest versatility, but it also has the highest cost as it is very expensive to have a gripper that will do everything. At the other end of the scale is the special gripper that is designed and made for a narrow, specific task. Its relative cost will be less than any other gripper, but its versatility will be very limited, if not zero. All grippers fit within the matrix of cost and versatility. It is up to the designer to determine what is wanted and at what cost.

A Swedish paper[26] discusses the economics of gripper design and determines, within the bounds of a generalized case, that the cost of assembly per unit increases more rapidly as a function of the number of parts requiring exchange of grippers than as a function of the number of parts themselves. Figure 38 shows the graph that substantiates this claim where it can be seen that for a given number of parts that the per unit cost increases in relation to the number of grippers used. Figure 39 shows the relative costs incurred for the use of a number of separate grippers to assembly a given product. The comparison is made on the assumption that the time taken to exchange the gripper is equal to the time taken to assemble a single part (8 sec in this case). The result is that if, for example, the number of gripper "exchanges" is reduced from five to three per product, then the cost saving per product varies between 0.8 and 1.0 Swedish krone. This means that the amount of effort that can be spent upon designing a more complex multifunctional gripper can be quantified from the reduction in processing time derived from the new design. The equations used for this assessment are:

$$C = (T_a/T_i)I + C_e + C_b$$
$$I = C_r + C_g * (n_g + 1) + C_p + n_p(C_i + C_p) + C_d$$
$$C_e = C_i\{(n_p * n_f * T_i * L) + T_0 + [T_i(1 - X) * L]\}$$

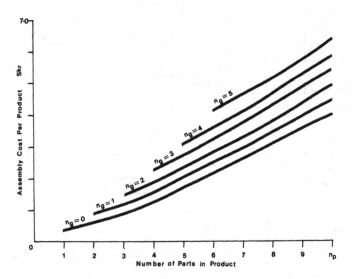

Figure 38. Assembly cost per product as a function of the number of parts in the product(s) and the number of gripper exchanges required. [Courtesy of IFS(Publications)Ltd., Bedford, England.]

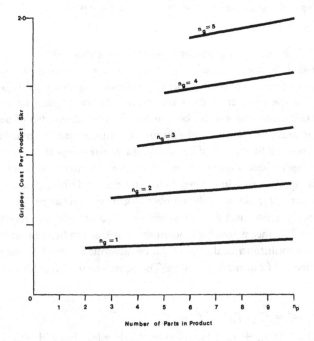

Figure 39. The cost of grippers per product assembled as a function of the number of parts in the product(s) and the number of gripper changes required. [Courtesy of IFS(Publications)Ltd., Bedford, England.]

where C is the mean production cost to assembly a product, C_d is the cost of sensors, guages, etc., C_e is the cost penalty of efficiency, C_f is the cost of feeding equipment per component, C_g is the cost per separate gripper, C_h is the cost of overhead (rents and operational costs), C_l is the cost of service personnel and operators, C_p is the cost of planning the system, C_r is the cost of the basic robot, C_s is the cost of storage for equipment, C_z is the cost of the automated system, I is the investment, L is the number of service personnel, n_f is the percentage of faulty parts, n_g is the number of parts requiring exchange of grippers, n_p is the number of parts in the assembled product, T_a is the mean assembly time per product, T_i is the idle time of the system caused by a faulty part, T_o is the time occupied by the system operator, T_t is the total available assembly system time, and X is the mean availability of the assembly system (percentage of total).

7.4.3. Economics of Task Times

Rogers[27] states that even the most economic robot system must be justified by comparison with existing assembly methods. Also, that the economics of a robot system tend to be more sensitive to nonfixed costs, such as tooling and arm speed, than they do to fixed costs, such as those associated with arm hardware. A cost equation describing the assembly per component (in the product) can be summarized as:

$$\frac{C}{n} = (t/2 + 2XT)\left(3W + \frac{W\{2R + N[C_c + nC_g + (ny + N_d)C_m]\}}{SQ}\right)$$

where C_c is the cost of the work carrier ($1,000), C_g is the cost of the gripper per component handled ($500), C_m is the cost of the manually loaded magazine ($500), N is the number of products (1), N_d is the number of design changes in the life of the assembly (0), Q is the equivalent cost of the operator in terms of capital equipment ($30,000), R is the cost of one robot ($40,000), S is the number of shifts (3), T is the system down time due to defective parts (30 s), W is the operators rate ($0.002 per second), X is the ratio of faulty parts to acceptable parts (0.01), n is the number of parts in the assembly (13), t is the mean time of the assembly for one part (5 s), and y is the number of styles for each product (1).

To illustrate the effects of fixed and nonfixed costs, the bracketed values can be substituted into the equation. Further, if the cost of the robot is halved the effect on the per unit price is insignificant. But if the robot "speed" is doubled—which has the effect of halving the assembly time—then the per unit price is reduced by almost 40%. Therefore, it can be seen that speed of operation is a more sensitive parameter for assembly cost. Consequently, it makes the estimation of cycle times a very important factor in the evaluation of alternative systems.

There are three methods of estimating the time to perform a task by a robotic system. They are all based upon the fact that robot performance is predictable and repeatable and that when the dynamics and control algorithms of a particular robot are known, then it is a straightforward procedure to calculate assembly times. The first is a quick, straightforward, approximate method that uses extrapolated data from experimental results.

A given robot has an average time for performing an average assembly task. If this average time is t_c and the number of parts and tools used in the assembly is n, then the time required for assembly t_a is

$$t_a = n * t_c$$

For a system with a variable number of grippers and robots, and assuming that any gripper is not changed more than once, the number of hand changes H can be calculated thus

$$H = h - N$$

where h is the number of grippers and N is the number of robots. Therefore, the time taken for gripper changes t_b is

$$t_b = (h - N)t_c$$

The approximate total time required for assembly, including gripper changes, will be the sum of t_a and t_b divided by the number of robots, assuming perfect efficiency in balancing a multiarm system. Experience shows that a 10% increase in overall time can be expected for each additional arm added. The total assembly time T can therefore be expressed as

$$T = \left(\frac{t_a + t_b}{N}\right) + (N - 1)0.1\left(\frac{t_a + t_b}{N}\right)$$
$$= \left(\frac{0.9}{N} + 0.1\right) * (n + h - Nt_c)$$

As the value for t_c is determined from averaging the time for many complete cycles, it is reasonable that inaccuracies will arise in the calculated value of T when the number of assemblies per arm (n/N) is small. It is estimated that where the value of n/N is greater than 5, the calculated value of T, by this method, will be within 20% of the actual time.

A more complex system for determining cycle times (when the ratio of n/N is small or when the assembly sequence deviates significantly from a series of complete cycles) requires a finer breakdown of the assembly cycle for evaluation, whereas the simpler system assumes a task time (5 sec) irrespective of the task—which, for a single gripper/robot combination, can be reduced to

$$T = 1.0\ nt_c$$

The more sophisticated system defines the basic task time as $8t$ which can be modified in two ways depending upon the complexity or length of motions required to complete the task.

The pick-up, discharge, or use of any object requires an approach and a retreat of the robot's end effector from a point in the near vicinity of the object. If the approach/retreat requires only one point-to-point move, no additional time will be added to the value of $8t$. For any action that requires a slow step, a taught time delay, a continuous path step, or more than one point-to-point step, then an additional time t will be required. The maximum additional time for any complete assembly cycle will be $4t$ (one for each approach and retreat).

The time required for the robot to move between one retreat and the next advance varies as a function of the distance travelled and the number of steps required. For an action of one set of combined arm motions, where the longest major motion is less than 0.3 m, or of two sets requiring only minor motions that do not exceed half the axis length with no stop at any intermediate point—a line equal to $2t$ will be deducted. For any complete assembly cycle, a maximum time deduction of $4t$ can be realized. Therefore, for a task time of 0.5 s the total time for completed assembly can vary between the extremes shown below

$$\text{Min} \qquad 8t - 2(2t) = 2 \text{ sec} \quad \text{to} \quad 8t + 4(t) = 6 \text{ sec} \qquad \text{Max}$$

There will be assembly sequences where the arrangement does not require a complete assembly cycle for each part. For example, one gripper could unload several molded parts from a cavity mold before returning to the assembly fixture to deposit them. A half-task is therefore defined as moving from the completion of one retreat through the completion of the next retreat and is considered equal to $4t$. This time can be modified using the same rules as before with the maximum deviations allowed being limited to half of the previous values. Therefore, the time to complete a half-task will range between the extremes shown below

$$\text{Min} \qquad 4t - 1(2t) = 1 \text{ sec} \quad \text{to} \quad 4t + 2(t) = 3 \text{ sec} \qquad \text{Max}$$

Although this intermediate time system requires a strong background in programming and a complete familiarity with the control system in use, it does result in a more accurate time estimate for each portion of the assembly cycle and permits balancing of a multiarm system.

An example of the use of the intermediate time system is given in Table 19. It was used to determine the assembly time for assembling a transmission governor by means of a double Unimate 6000 robot. The intermediate time system predicted an overall time of 30 sec. The simple system predicted 35.75 sec and the actual time was 31 sec. The value of 35.75 sec is obtained by using $n = 13$ (from the right-hand column on Table 19, which is the largest number of actions), $N = 2$, and $h = 2$.

Today's industrial robots often boast a life exceeding 40,000 hr, which is the

Table 19
Assembly Sequence with Time Estimates Determined by Intermediate Time System

Left arm	Time	Units	Right arm
SB off feeder and onto fixture $8t + t + t$	$10t$	$3t$	Insert two screws $4t - 2t + t$
		$6t$	Return screwdriver $4t + t + t$
LV off feeder $4t - 2t + t + t$	$3t$	$10t$	FA of fixture and onto exit chute $8t + t + t$
Align SB $4t - 2t + t + t$	$4t$		
Insert LV into SR $4t - 2t$	$2t$		
LV1 off feeder and insert into LV $8t - 2t - 2t + t + t + t + t$	$7t$	$4t$	LV3 off feeder $4t - 2t + t + t$
		$3t$	LV2 off feeder $4t - 2t + t$
SV off feeder and into SB $8t - 2t + t + t + t + t + t + t$	$10t$	$6t$	LB off feeder and onto fixture $8t - 2t$
		$4t$	Insert LV2 & LV3 in SB $4t - 2t + t + t$
SV1 off feeder and into SV $8t - 2t - 2t + t + t + t + t + t + t$	$8t$	$11t$	SV2 off feeder and into SB $8t + t + t + t + t$
Hold SV1 ($\equiv 3t$)		$6t$	Get screwdriver $4t + t + t$
SB off fixture and onto LB $8t - 2t + t + t + t + t$	$9t$		Wait ($\equiv 2t$)
Hold SB ($\equiv 4t$)		$5t$	Insert two screws $4t + t$

equivalent of 20 shift years. Consequently it may prove worthwhile to perform a detailed time analysis to accurately predict cycle times, especially when it is required to balance a multiarm system.

An efficiency figure for a multiarm system consisting of N robots with individual assembly times of T_i s and resulting in an overall cycle time of T (including waiting time) can be expressed

$$E = \left[\sum_{i=1}^{N} T_i / NT \right] 100$$

Using the values from Table 19, we have

$$E = \frac{26.5}{2 * 30} + \frac{29}{60} = 0.925 = 92.5\%$$

In developing multiarm systems, it is important to arrange the assembly sequence so as to minimize waiting time and thereby maximize the efficiency of the system.

This detailed time system is an expanded version of the previous method and requires a complete knowledge of the assembly station geometry including the relative positions of arm(s), tooling, fixtures, and parts, as well as knowledge of robot dynamics and a detailed description of the robot control algorithms.

The arm trajectories for robots can be point-to-point and/or continuous path. A generalized point-to-point step results in all articulations (that are involved) starting to move to their destination—usually in unison. Each actuation goes through an acceleration, slew, and deceleration period with each motion finishing quite independent to the others. Any number of these points can be linked together with the condition that all articulations are in true positional coincidence prior to proceeding to the next step. Alternatively, points can be taught such that the memory will update to the next step before position coincidence is met, thus creating a rounding effect which results in a significant reduction in time.

The times required to perform the various trajectories can be calculated using information describing arm dynamics and control parameters. A continuous path trajectory can be programmed such that the end effector follows a determined path in space over a wide range of velocities. The linear interpolation scheme used for continuous path steps, allows a simple calculation of step velocity by dividing the number of interpolations per step i by the frequency of interpolations f.

The calculation of the point-to-point steps is more difficult and requires a knowledge of acceleration rates and slew speeds for each articulation. A typical velocity profile for a point-to-point step is shown in Figure 40 and portrays a linear acceleration a followed by a slew velocity v_s (for steps long enough to saturate), and finishing with a linear deceleration a. The calculation of the time required to move distance S is

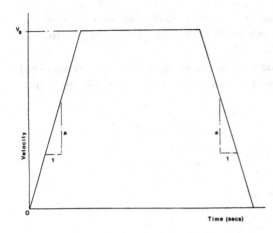

Figure 40. A velocity profile for a robot. It is often the case that the constant "slew" velocity is not reached before the robot has to decelerate to its terminal position for the action. [Courtesy of IFS (Publications) Ltd., Bedford, England.]

$$t_s = \frac{S}{v_s} + \frac{v_s}{a} \qquad S > \frac{(v_s)^2}{a}$$

The time required for a step of length S to be covered where the velocity saturation is not reached is

$$t_{ns} = \frac{2S}{a} \qquad S < \frac{(v_s)^2}{a}$$

Because of the independence of completion times for each of the articulations in a point-to-point move, a generalized motion consisting of up to six degrees of freedom requires the calculation of the time needed for the slowest articulation. Knowing the coordinates x, y, z, θ, β, ∂, of the points between which the motion will occur, the transformation matrix M of the robot enables the step length S_i of each articulation to be calculated. These values of S_i can then be submitted for S in the previous two equations to determine t_s or t_{ns}, respectively.

The time required for any generalized assembly cycle can be found by adding together all of the times required for arm motions under normal memory control, associated time delays, waiting time, and times required for arm motions under sensory feedback control.

The nature of part insertion using sensory servoing (e.g., force and vision) results in a variance of assembly time. Waiting time, however, can be accounted for and will be defined as that time where any device (including another robot) causes the robot to be stationary. The time can be attributed to external tooling such as part feeders, index tables, etc., or two-handed coordinated tasks where one robot may be used as a holding fixture. An additional waiting time is not added to the individual assembly cycle but to the overall assembly sequence due to the inability to perfectly balance a multiarm system. Two types of realizable time delays are the clamp time delay and the programmed time delay. The former is preset and allows for gripper opening and closing time, the latter is adjustable to allow for interaction with external equipment.

The calculation of time required for a complete assembly sequence using this detailed system is tedious and needs a detailed layout of the assembly station, including the design of all the tooling and other ancillary equipment. Also, a complete choreography of arm movements for the complete assembly cycle for each part that is used is needed. Using an assembly precedence diagram, a preferred assembly sequence can be developed for a system using a single robot which can then be analyzed for the overall cycle time.

Even before the assembly sequence can be arranged for a multiarm system, the work load for each robot must be carefully balanced to reduce costly waiting time. This requires the predicted time for each complete assembly cycle or part thereof to be performed by each arm. It is also necessary to take into account any variation of cycle times due to down time through faulty components. These items, while incapable of being processed, hold up the system output until they can be cleared. The one exception to this is a very sophisticated system whereby each part is checked before the next action is initiated. If the component is found to be unacceptable, then it is rejected and the action repeated until a suitable part is presented.

7.5. Robot Safety

"Robot murders worker" was the headline in December 1981[28] when the media reported the "beating to death" of a Japanese maintenance worker by a robot at Kawasaki Heavy Industries near Tokyo. What happened is subject to conjecture. The reason stated in the press was that the human had entered the robot's working space by "leaping" over the barriers, whereas if he had gone through the proper access gates and entry procedure, then the robot would have been automatically powered down and consequently harmless.

According to the Health and Safety Executive in London, the individual was supervising the progress of work on a production line and—while there were no witnesses—it is thought that one of the machines stopped as a result of a door at the front of the machine failing to open. The human switched the machine to manual and disengaged the coupling switch to the robot. Having cleared the blockage, which allowed the door to open, the coupling switch was engaged. This caused the robot to extend its arm to remove a component from the machine and the operator was crushed by this movement.

Unfortunately, automation in general and robots in particular suffer from bad PR at the hands of the media, who often use distorted headlines to sell newspapers and television time. However, what is even more unfortunate is that there is not (as yet) any definitive form of industrial codes of practice to specify guarding systems that should prevent any human or robot from harming one another (Asimov's laws excluded). The usual compromise is that the guarding should comply, as far as possible, with that specified for automated machines and what the local safety officer feels is needed. Consequently, until such a time as rules with teeth in them exist, the risk of injury to a human by a robot remains a serious problem.

Barrett *et al.*[29] states that the degree of guarding depends upon the risks involved. The risks are determined by the frequency of access to the danger area, and the severity of injury is related to the method of operation, the likely need for access, the action of the parts safeguarded by interlocks, and the characteristics of the system.

There are several methods of interlocking the control system. There is control interlocking, power interlocking, control with back-up, and dual circuit interlocking. Control interlocking is considered acceptable only for normal interlocking situations, while the other methods of interlocking are considered acceptable for high-risk situations. The essential criterion is to establish that the integrity of the electrical system is satisfactory for the risk involved.

Consequently, and this is particularly important in the context of robotic installations, the robot might not be considered dangerous when operating as the designer intended. It may, however, be considered dangerous in the event of a control system malfunction. Consequently, if no other safeguards are proposed and sole reliance is being placed on integrity of the robot control system, the precautions built into the control system against a malfunction occurring should be of comparative integrity to that which would be applied to an interlocked guard for the same risks in question. From this it can be simplistically stated that a reasonable level of electrical safety integrity applies when:

   (i) the safeguarding is reliant solely on the robot control integrity, or
   (ii) the safeguarding is reliant solely on traditional safeguarding methods, or
   (iii) the safeguarding is dependent upon both the robot control system integrity *and* traditional safeguarding methods.

The robotic system is considered in three possible principle modes: normal working, programming, and maintenance. For each mode there are two possible behaviors:

Normal working: Designed or aberrant behavior
Programming: Designed or aberrant behavior
Maintenance*: Designed or aberrant behavior

*If powered motion is required during procedure.

The term "designed behavior" is self-explanatory and means that the system is operating in the intended manner to which it was designed. The term "aberrant behavior" refers to any uncovenanted movement of the system caused by a malfunction of the control system. For example, the robot may behave aberrantly because of electrical interference or due to an error that has been introduced into the program. The error may have been inherent in the program as conceived or may have been introduced into the program in its digital form as a result of some transient fault or disturbance, e.g., a glitch.

While the basis of control of a robot, namely, the programmable electronic system, has tremendous potential for flexibility, it also has the potential for malfunctioning. This malfunction will possibly be in an unpredictable manner and probably to a higher degree than systems operating on conventional (nonprogrammable) lines. The failure modes associated with programmable electronic systems are complex and any analysis to quantify the overall safety integrity is also likely to be very complex. However, if safety is involved and is dependent upon a programmable electronic system, then some form of safety integrity assessment must be carried out and the safety integrity assured, or it must be assumed that certain malfunctions will occur and the safeguards should be adopted based upon this premise.

Figures 41 through 46 are an algorithm for examining the system to discover the

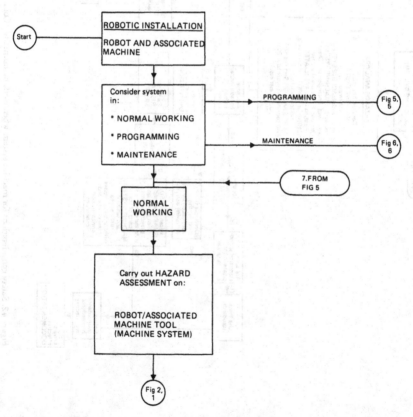

Figure 41. Basic installation comprising: robot/associated machine. (Courtesy of Health and Safety Executive, London. Crown Copyright.)

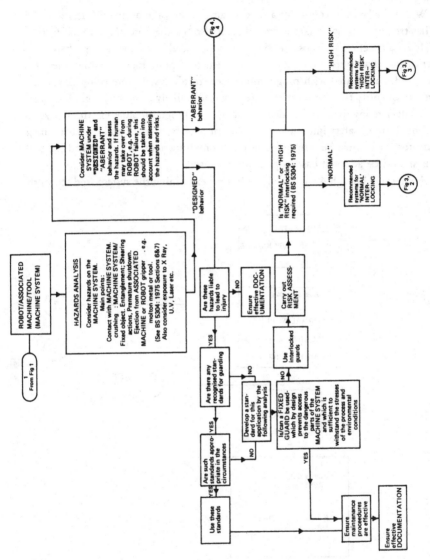

Figure 42. Safety considerations for machine system when working normally. (Courtesy of Health and Safety Executive, London. Crown Copyright.)

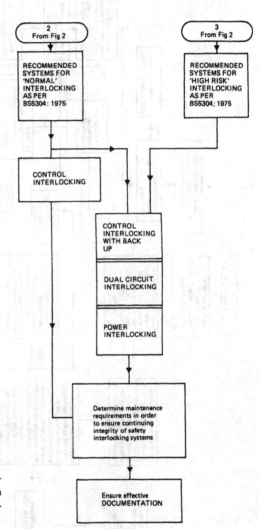

Figure 43. Recommended electrical interlocking systems for "normal" and "high risk." (Courtesy of Health and Safety Executive. Crown Copyright.)

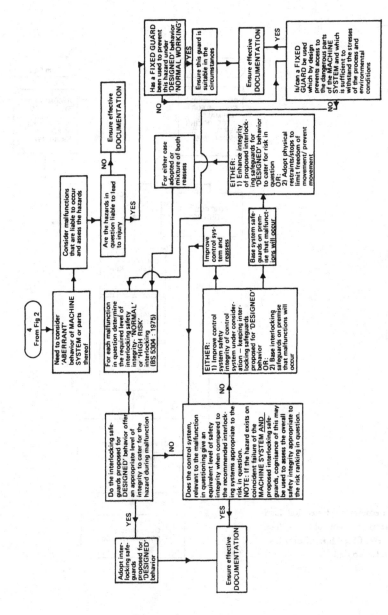

Figure 44. Safety considerations for machine system when "normal working," but subject to "aberrant" behavior. (Courtesy of Health and Safety Executive. Crown Copyright.)

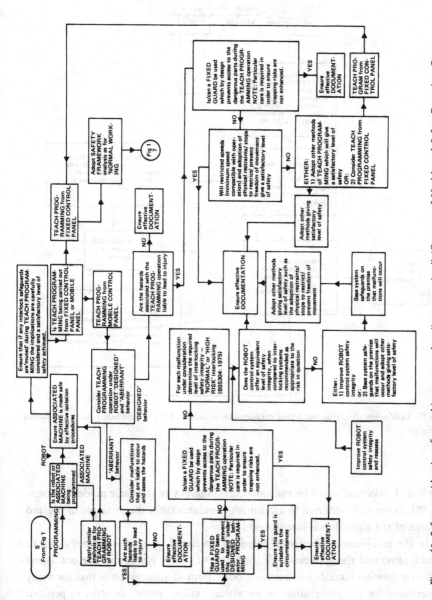

Figure 45. Safety considerations for machine system when "programming." (Courtesy of Health and Safety Executive. Crown Copyright.)

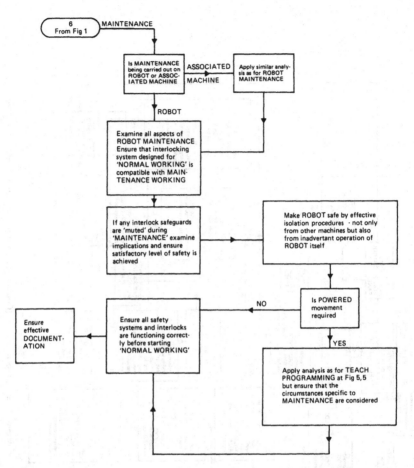

Figure 46. Basic safety considerations during maintenance. (Courtesy of Health and Safety Executive. Crown Copyright.)

hazards and risks that could be encountered during any of the three modes of operation. While the codes referred to are British Standards (BS), the principle of use and the value of the algorithm is international.

In the end—until definitive regulations are issued by law—it comes down to common sense and the application of the "what if . . . ?" syndrome. The one aspect of robotics that distinguishes it from any other form of automation is that for many it is necessary for the programmer to be within "killing range" in order to perform the programming tasks. It is acknowledged that robot motions during programming are usually only a fraction of their functional operating speeds *but* malfunctions do happen *and* it is possible for the inexperienced operator to get into trouble through overinterest in the process and ignorance of what is happening to the robot's appendages and associated equipment, i.e., the programmer could be in a position whereby the robot movement from one point to another hits the programmer and causes a random instruction from the programmer (in panic) that activates the robot, *et cetera, et cetera.*

Machine Senses

The replacement of a human by a robot is based upon a series of socio-economic criteria. The effective use of a robot necessitates that it is provided with information about its environment so that it does *not* perform its tasks irrespective of what is happening within the work station. The most primitive robot system will, for example, allow the painting sequence for an object to be performed whether the object is there or not and/or whether or not it is the correct object. The reason that this can occur is that the particular systems that permit this to happen are time-based sequence systems without any interlocks to ensure that the "environment" is correct for the task performance.

Humans (usually) are equipped with the senses of sight, smell, hearing, touch, and taste. These senses enable them to make direct observations and indirect interpretations of their working environment, so that task related decisions are easily and rapidly made. Therefore, in order for a robot to function in tasks that require "human awareness," they must be equipped with the machine equivalents of the human senses, so that they can examine and monitor their environment, as well as responding to any stimuli.

Unfortunately, such a range of sensory perception is as yet beyond the capacity of industrial robots, albeit robots have been developed with limited equivalents of a few of these senses. For instance, probes, proximity switches, and microswitches have for years been used to indicate, by tactile or other means, that an object is present. The problem is that these systems operate only in two states, they are either Off or On, there being no capability to distinguish between Here, Near, or Not Here.

8.1. Touch

Another example of these primitive systems is the type used for allowing an assembly to take place, between, for example, a peg and a hole. If the two components are not correctly aligned, then, being a "rigid" system, a wedging or jamming condition occurs. If a flexible "compliance" device is placed between the machine presenting the peg and the gripper that is holding the peg (simplistically it forms a part of the gripper), then again if assembly does not occur naturally a collision state results. However, in this case, the collision causes rotational and lateral forces to be generated, the magnitude and direction of which, in turn, generate the corrective motions in the passive device so that assembly results. In active force feedback compliance systems, the generated forces activate servos to achieve assembly. The net result with both systems is that the peg enters the hole with the minimum of wasted time. These compliance systems are commercially available and will function with chamfered or chamferless "pegs" and "holes" as well with a range of clearances. Figure 47 explains the difference between wedging and jamming, while Figure 48 and 49 show the relationships that permit the RCC (a passive compliance device) shown in Figure 50 to function. The equivalent diagrams for an active compliance device are shown in Figure 51, and the range of applicability and the strategy options for both style of compliance systems are given in Figure 52.

8.2. Smell

Robots have recently acquired the sense of smell. While analyzers have existed for some time as a means of identifying substances through chemical reactions, there has not existed before a mobile device that could make decisions based upon a "sniffing" action. The robots that have been credited with first acquiring this sense are two Trallfa units on the assembly line at Austin-Rover of Cowley in the United Kingdom, where they are used to "leak test" automobiles.

The automobiles are leak tested to check for water ingress paths on finished automobiles prior to their coming off the line. The traditional method of using high-pressure water tunnels has many disadvantages, such as damage to internal fittings—if there is a leak—as well as initiating a rusting action. Neither is the traditional

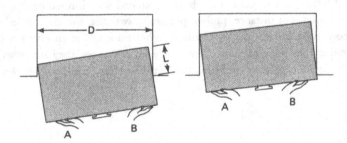

Figure 47. Difference between wedging and jamming was clarified during the development of compliant-gripper mechanisms. When, for example, a bureau drawer becomes wedged (left), it is literally locked. Any further application of force will deform the drawer or the bureau or both. Theory shows that wedging arises when the drawer is inserted at such an off angle that the ratio of L/D is less than the coefficient of friction (μ) when two-point contact first occurs. The only remedy is to pull the drawer out and start again. If, however, the ratio L/D is larger than μ at the time of initial two-point contact (right), wedging cannot result, although further movement can be impeded by jamming. The remedy is to break the two-point contact by pushing at A, thereby changing the direction of both the applied force and the applied moment.

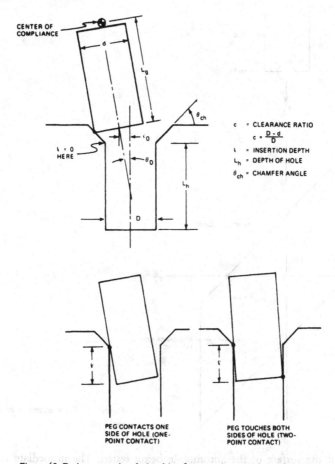

CENTER OF
COMPLIANCE

θ_{ch}

$\iota = 0$
HERE

c = CLEARANCE RATIO
$c = \dfrac{D-d}{D}$
ι = INSERTION DEPTH
L_h = DEPTH OF HOLE
θ_{ch} = CHAMFER ANGLE

PEG CONTACTS ONE
SIDE OF HOLE (ONE-
POINT CONTACT)

PEG TOUCHES BOTH
SIDES OF HOLE (TWO-
POINT CONTACT)

Figure 48. Basic geometric relationships for a passive compliance device.

system very accurate, since it cannot simulate the fine emulsification of water droplets as produced when driving along a damp road, or when the car has been standing in the rain for several days.

After a number of investigations the decision was made to reverse the direction of the leak identification. In the new system, the trimmed body is "seeded" through the gearshift hole with 0.5 l of helium under a slight pressure. Two robots equipped with "noses" are then used to sniff out gas leaks along a 60-m path around doors, windows, trunk, and seams. The inspection time is 165 sec and the rate of inspection is 18 cars per hour. Accuracy and consistency are ensured by the use of a "control" leakage through a pad fitted over the door mirror mounting hole; also, the special nose that contains the helium sensor is shrouded with a dense air curtain. The robots first check themselves against the control leakage and then present their noses within

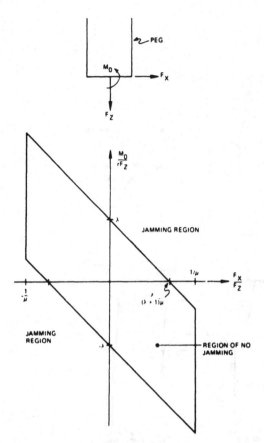

Figure 49. Force moment relationships to avoid jamming.

25 mm of the surface of the automobile being tested. The immediate atmosphere around the standard gas analyzer probe fitted inside the nose is enclosed by a circular and continuous air curtain, which allows the sensor to measure very precisely the strength and location of each gas leak.

The robot noses presently used are about 150 mm in diameter, but work is going ahead to design and manufacture 50-mm-diam units that will allow access to more restricted areas of the car body. The units are finding not only more leak paths than did the water tunnel, but they are highlighting routes of fume ingress and faults with the door seals that, while unlikely to leak water, will create undue wind noise at speed.

8.3. Sight

The sense that accommodates the most information is that of vision, especially steroscopic vision. The primitive "vision" systems used in industry are those of pho-

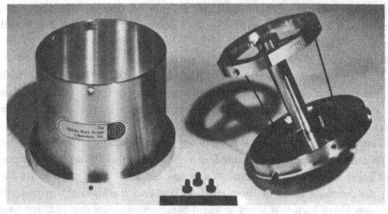

Figure 50. Model 4B remote center compliance. Designed to provide a lighter weight (1 lb. or .45 kg mass) unit capable of easier repair and lower cost. The focal length and elastic parameters are nominally the same as the Model 4A.

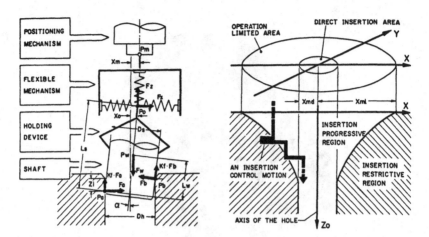

Figure 51. The upper diagram shows a two-dimensional representation of an insertion operation by an active compliance device, while the lower diagram shows the insertion characteristics for the same device.

tocells which, as with the primitive tactile systems, only have two states—Yes/No or On/Off. Likewise, the information obtained from this one bit of information only permits the system to respond with the same limited two state option—Go/Stay or On/Off.

Fortunately, vision systems are now being developed and used (in restricted applications) that can detect differences in color density and texture through their ability to distinguish color in 256 shades of gray (possibly even more than the human eye). This means that it is now possible to identify objects—not just by presence, but by recognition—such that the object's shape, attitude, orientation, and certain physical properties, such as center of gravity and least moment of inertia, are determined in real time.

Simplistically, these systems work as a result of the image being broken down into a matrix of squares, typically 125 X 125; 250 X 250; or 500 X 500. Each of these squares—termed a pixel (picture element)—is assessed individually for classification into the appropriate gray level. The simplest choice is binary one, where each pixel is recognized as being either black or white. After classification, the information is evaluated by computer and the appropriate action is initiated. Once the image is stored, it can be manipulated, presented, and analyzed so that the desired information is most easily extracted. For instance, an image can be changed from full color to binary; the threshold levels can be filtered, so that only the highlights remain; there are low-pass filters; high-pass filters; gradient operators; template matches; line finders; contour followers; feature extractors, etc.—all of which can be applied singly or in combination to give the desired result.

There are many potential applications for the use of vision systems within the FAS. The applications can be divided into two main areas. The first is the use of visual

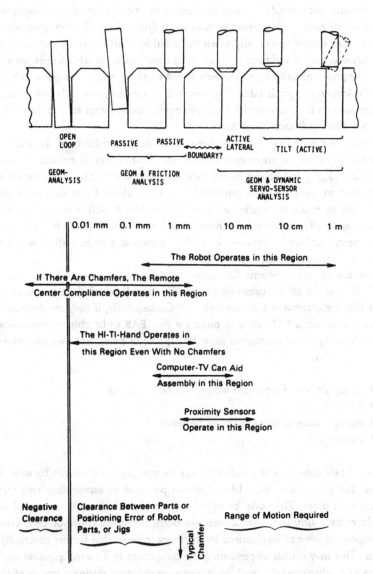

Figure 52. The upper diagram shows the error regions and strategic options for compliance devices, while the lower diagram indicates the range of applicability of the compliance systems and assembly techniques.

information as an essential part of the control of a robot. The second is the aiding or replacing of humans in inspection tasks.

Presently, the control of a robot depends upon its own internal positioning system and the use of passive compliance devices when fitting close tolerance parts together. However, this system works only when the position of all the equipment within the work station is known and fixed—for specified time frames. In other words, for a given action, the position, attitude, and orientation of every jig, tool, work-piece, robot, and gripper is known and predictable. The system is therefore dynamically inflexible, since it cannot account for uncertainties within its environment except for the simple binary Here/Not Here indicators of presence.

Vision sensing offers a considerable increase in the flexibility of the system and allows adaptation to the uncertainties within the environment. It reduces the amount of expensive jigging and programming time used to restructure the work station when new component types are to be processed. It is acknowledged that nonvision systems can be used to make the workplace flexible, but this flexibility is limited, since a nonvision system has either a finite number of options before it "stops" or there is a need for search-and-find routine—like a blind person in a maze—all of which takes time to process, understand, and respond to.

The use of vision systems for robots in FASs, can be separated into four main areas.[30] While not all are presently viable, they are all practical problems and conditions that are encountered within industry. Consequently, if they are deemed to be encountered within a FAS, then in order for that FAS to be able to function at its highest efficiency, these conditions must be capable of being resolved automatically. The four areas are:

- manipulation of separated workpieces on conveyors
- bin picking
- manipulation in manufacturing processes
- assembly

The use of belt conveyors is a common way to transport components between work stations. The components may be in random positions or orientations and may be touching each other. They may be piled in layers, but they can be separated into a single layer by a simple passive device,—a beam across the conveyor belt. Usually, the components have to be acquired by the robot in order that further processing can be done. This may consist of presenting the component in a correct position and orientation to another machine, packing it into a container, or sorting a stream of mixed components.

If the components do not occlude each other, it is feasible to apply present-day vision techniques to identify the component and so determine which stable state it is in. It is also possible to determine its position and orientation as well as calculating suitable gripping points. In order to achieve this information, it is imperative that a high-contrast image of the component is obtained so that its outline and other features can be determined. A 128 × 128 image, together with a mini- or microcomputer, can perform the required sensing for a wide range of such tasks. Current software is adequate for acquiring and processing the image within 0.5 sec, although by adding special preprocessing hardware, the time can be reduced to 200 msec.

If the components are hung from an overhead conveyor, then more complex processing must be done in order to determine the position and orientation of a particular component. The components are usually crudely supported and, consequently, 3D information is required to provide the necessary data. There are two approaches to this situation: the first is to ensure that the components are rigidly supported, hence guaranteeing a unique position and orientation in space, as well as eliminating sway and rotation as much as possible. Secondly, 3D information may be obtained by range sensing or by structured light.

The first solution is not usually economically justified, especially in the case of small batch production where customization of handling equipment would be totally nonsensical, also the recovery of 3D information from 2D images still requires much more research. Thus, the use of machine vision is not yet applicable to other than the simplest recognition of routine components on overhead conveyors.

Another common method of transportation and storage of components within an FAS is the use of containers or bins. The problems of dealing with bins of components is similar to that of dealing with components on conveyors, namely, that the components have to be separated before subsequent processing can be conducted. The use of a vision system to recognize individual components stored in a bin is a difficult problem, since the components are often jumbled together in random positions and orientations. They may be interlocked, with no clear unobstructed view of each component. In addition, each component will tend to cast shadows over its neighbors within the bin.

While the "bin-picking problem" has been and still is a major research project, there is not yet a viable vision system controlled bin-picking system used in industry. It is considered by many groups, that the problem is of academic value only, since there are many less complex answers to the removal of components from bins. If they must be presented jumbled in a bin, then a relatively crude mechanical device will search and pick out individual components. It will also drop them onto a conveyor or present them to another device for correct identification, reorientation, and presentation to the processing equipment.

If the components are arranged within the bin in a defined pattern—as advocated by BRSL[2]—then the task of "bin picking" becomes easier. If the components do not occlude each other and there are no shadows, bin picking may be performed with present day technology. The task is made easier by the use of specially tailored backgrounds that provide a contrast to the components. This is often feasible since the components may be placed in special "egg crate" type compartments so as to protect them from damage and to separate them from their neighbors. These compartments can be designed to be of a contrasting color to that of the components, or printed with a fixed geometric pattern offering a contrasting pattern to that of the component.

A number of the fixing and forming processes within the FAS will require that the tooling follow an edge or seam on a give component. If the edges or seams vary in shape and size due to the vagaries of their manufacture, then a device is needed that will control the tool so that it follows the contours of the edge or seam. The use of a vision system to follow the joint line in robotic welding applications is now well established, whereby the welding gun is weaved to compensate for the waviness of the joint line. The application of this technology to the requirements of the FAS are found in liquid gasketing, laser machining and welding, and the application of semi-liquid sealants and the deburring of components.

When a person performs an assembly action, the parts are oriented by tactile and visual senses. If a component is obviously asymmetrical and the operator has been properly trained—then vision is not essential to achieve assembly. This is well illustrated by the capabilities of blind persons within a known or predictable situation. It is often the case that many of the components used within a product are similar—e.g., screws of different sizes or shafts of slightly different diameters—and that quick and correct assembly requires that the correct parts are assembled in the correct place and sequence.

It might also be the case that a different component requires a different process to achieve assembly. Also, damaged components are often more readily detected by visual means than by the use of other (machine) senses. It is therefore, practical for components to be presented roughly at a work station, where a vision system will scan and correlate the necessary information. The control computer will then instruct the robot to place any component in the correct attitude—many components have several stable resting positions—for assembly to take place. It will then pick up the components to be fitted and orientate the "mobile" parts to suit the orientation of the "static" items. The final assembly will be achieved by passive compliance devices and, if necessary, by a search-and-find vision routine to accommodate any other loose components.

A valid criticism is that vision for picking and assembly tasks alone is a good academic process that really has no use in industry, since there are simpler, less costly ways of ensuring correct attitude and orientation as well as achieving assembly. This would be true if it were not for the hidden benefits that vision can and does offer to the automated system. Firstly, it can accommodate a slightly stochastic work place, where everything does not have to be known to the nth degree. Secondly, it can be used by the robot as a navigation aid to avoid collisions with other mobile structures within the environment. Thirdly, it is the most comprehensive sense, giving the most information per unit of time. Lastly, it can be used to inspect the process as it is being conducted, and it can inspect the result of the process to check for completion, errors, and/or potential problems—or instance, sealant spreading into an area normally required to be kept clear by the process.

With any vision system functioning within an FAS, there is a need for a real time response. This means manipulating a mass of data in a very short period of time. Mass production inspection rates require accept/reject rates of 10 to 20 items per sec, pick-and-place operations may require part recognition at the rate of one every 2 sec, yet a typical image of 128 × 128 pixels consists of 16 K bytes if 256 gray levels are used, or 16 K bits if only a binary image is used. Either way, there is an enormous amount of information to be processed.

There are three approaches to obtaining a real time response. First, the software processing times may be reduced by cutting down the amount of image to be processed. This can be done by the use of software windowing so that only a subrectangle of the picture is analyzed. Alternatively, only selected isolated points may be analyzed. In both cases the portion of the image used for analysis is easily controlled by the program. The use of binary images, as has already been indicated, reduces the amount of information to be processed, although it limits what processing can be done.

The second approach to the problem is to perform all or some of the processing in hardware. In the case of analyzing the position and orientation of parts, and also in part inspection, hardware systems have been devised that perform part or all of the image analysis. Many systems perform thresholding by hardware, though not all of them allow software control of the threshold. Experimental systems have claimed processing rates of 125 parts per sec.

The third solution to this problem of real time response is to combine software techniques with fast special purpose parallel hardware. The hardware may consist of many small bit oriented processors, one for each pixel, or it may consist of a smaller number of microprocessors, each dealing with a portion of the image. The typical example are the distributed array processors, which when connected and combined with several neighbors perform very fast image processing.

Any vision system that is used within the FAS should be of modular design so that it is possible to expand the system when new vision processing techniques become available. The use of both hardware and software modules covering the entire range of available (applicable) techniques means that the system could be capable of handling a wide range of applications, with each application being performed by a subset of the available modules.

In order to determine the correct configuration for a given application, a certain amount of experimental work may have to be performed to achieve optimal results. This involves consideration of the best form of lighting and adjustment of the environment to obtain a suitable image. It also involves consideration of the processing requirements and the time taken to analyze the image. This experimentation is best conducted interactively using a range of different algorithms and lighting conditions, which requires a development system with a convenient user command interface. The final arrangement will inevitably be a compromise between what would be preferred for ideal response and what is suitable for the work station and the real world industrial application.

The results of any processing by a vision system must be made available to the environment in which it is functioning. If the environment includes a human (as supervisor), then the output merely has to be displayed on a monitor. This output can be mixed with graphics from the processor, if extra information is required by the human supervisor.

If the vision system is replacing a human in performing a task, then the result of the processing of information must be used to continue the operations being performed. This could be a signal sent to a diversion device so as to remove faulty components or worn tooling from the work place, or to stop a process and advise the supervisor.

If the vision system is controlling a robot, two possible situations occur. The first is when the system is used to identify objects and their position in order to determine the grip position for the robot. In this case, the output of the vision system must be transferred to the program that determines the required movements so as to pick up the object. In other words, the computer robot program reacts to conditional inputs. Once the vision system has analyzed the scene, it need do no further work, since the control of the movement of the robot is performed without the use of the vision system.

The second situation is where the vision system is actively controlling the path that the robot follows to perform a given task. Here the vision system continually analyzes the scene for as long as the manipulator is being moved, starting from the time that the robot is given instructions to begin a task—conditional on it being possible. As the arm moves, it receives information that enables course corrections to be made until the task is complete.

Before the visual image can be processed by the computer, it must first be converted into a form that is compatible with the processor. This conversion is divided into three parts, the recording of the scene, the quantification of an analog scene into digital values, and the transfer into the processor. There are two forms of imaging devices that are compatible with industrial usage requiring real time analysis. These are solid state cameras and vidicon cameras. Other devices such as microdensitometers and flying spot scanners, although offering extremely high precision are not able to record a scene directly and must be used with a transparency or photograph for analysis or machine guidance.

8.3.1. Vision Equipment

The vidicon camera is relatively inexpensive, readily available, and fast. It allows direct analysis of the scene from the image that is focused on the end of the vidicon tube. The end of the tube is coated with photoconductive material which generates a photoconductivity pattern that matches the pattern of brightness. An independent, finely focused electron beam scans the rear surface of the photoconductive target and translates the pattern into another pattern of varying voltage that is proportional to the brightness.

There are two types of solid state cameras: photodiode and charge-coupled devices (CCD) cameras. Photodiode cameras focus the image onto a line or array of photosensitive diodes, which either conduct or do not conduct, dependent on whether or not the light value is above the preset threshold. The resultant binary image may be retrieved by scanning the array or by random access. CCD cameras also focus the image onto a line or array of photosensitive devices. In this case, however, the light falling onto the elements cause a charge to be built-up at a rate proportional to the brightness level. The image is scanned by moving the pockets of "charge" into a shift register and presenting the charge pockets one at a time to the digitizing stage. The CCD camera is more expensive than the vidicon, if a 2D array is required. However, the price of the CCDs will fall since the technology is still relatively new and is based upon LSI, whereas the price of vidicon cameras will not fall to any great extent.

For industrial purposes, CCD cameras are more suitable since they are robust, small, and require little power to operate. They are also more accurate than the vidicon units since their sensing elements are precisely known, in comparison to the uncertainties of the scanning principle. Hence, they are used for applications where measurement accuracy is important. Finally, CCD imaging chips are available in various sizes; for example, 100×100, 488×380, 1024×1, etc. If relatively low resolution is required, then a chip with fewer elements will allow a faster scanning rate, whereas a vidicon camera is restricted to a standard scan rate for output to a TV monitor.

After the image has been formed it must be quantified into a form suitable for input to a processor. The quantification occurs in two stages. First, the continuous image must be sampled to give a finite number of pixels. This is obtained by the digitizing of the spatial coordinates. Second, the gray level values of the image at each pixel must be digitized, since the dimensions of the array of pixels and the number of gray levels chosen clearly effects the quality of the image (a picture of monochrome TV quality requires at least 512×512 pixels with 128 gray levels). Many industrial applications using the same number of gray levels require only 128×128 pixels, but often an ideal picture consists of a binary image of two gray levels, one representing the object and the other the background. However, this binary image is nearly always obtained by thresholding a 64 or 128 gray level image, since the value of the threshold may have to be adjusted from picture to picture.

Once the image has been digitized, the data must be made accessible to the processor. There are three ways in which this can be done. First, each pixel may be presented at a parallel interface and transferred into memory by a program that reads each pixel. Second, the data can be input directly into the memory by a DMA channel. Third, the video system can input the data into a bank of memory contained in the video system itself, called a frame store. This last method has a number of advantages over the first two. These are that the data acquisition rate can be faster since the memory may be accessed directly; also, the use of the memory in the frame store for each picture storage frees the memory in the processor, since the image is in the frame store memory. Last, the frame store can be used to provide output to a TV monitor, thus displaying the current digitized image. Images can be processed, enhanced, and overlaid with text in the frame store, with the monitor continually displaying the result of the processing.

8.3.2. Lighting Equipment

The lighting of the scene to be analyzed has considerable bearing upon the feasibility and the results obtained. There are three basic considerations: (i) the use of lighting to obtain a high contrast image, (ii) the use of structured lighting, and (iii) the use of special lighting effects.

Lighting may be used to simplify the vision analysis of a task by preprocessing the scene into a high-contrast image suitable for thresholding into a binary image. Suitable effects may be obtained from using any one of a number of lighting techniques and a careful choice of lighting:

1. The use of daylight or normal photographic lighting, will in many cases suffice to illuminate the scene.

2. The objects to be analyzed may be backlit by placing them on a translucent platform of belt, illuminated from behind. This technique is good for producing black silhouettes on a white background. It is particularly useful for analyzing objects with holes, whereas surface detail will not be easily detectable.

3. Shadows may be reduced by using multiple light sources, and enhancing through directional lighting. Highlights can be reduced by oblique lighting and enhanced by illumination from a direction near to the camera.

4. If the object is moving fast enough to blur the image, a flashgun or strobe can be used to freeze the movement. Alternatively, the camera can be provided with a shutter.

5. Different parts of the light spectrum may be used to enhance the contrast of the scene. For example, red objects on a dark background may be enhanced by the use of a red filter. Special tubes are available for vidicon cameras which have different spectral sensitivity curves. Infrared sensitive vidicons are used in several applications. They are used to record temperature variations in steel furnace linings to detect breaks in the lining. They are also used to discriminate between grasshoppers and green beans in a packing plant—chlorophyll has a strong reflection band in the near infrared, grasshoppers do not. In an application with an arc welding torch, laser light is used to allow discrimination between the laser light and the arc welding torch, so that control of the bead could be gained.

6. Objects or background can be painted to enhance the contrast. It is usually more convenient to paint the background, and unusual application of this was the use of red fluorescent paint which, when the scene was illuminated with UV light, fluoresces while the object does not.

The use of structured light is another method of preprocessing the image, especially if 3D information is required, rather than a 2D silhouette. Structured lighting consists of the projection of a light pattern with a known geometric property. For example, a line of light generating a point on the object, a plane (sheet) of light generating a curve or line on an object, or a grid pattern generating a corresponding grid on an object. These lighting techniques are usually used to obtain 3D information.

Special lighting effects are used when, for example, the requirement is to inspect bottles for flaws. The bottles are passed between crossed polarizers under illumination by strong monochromatic light. A pattern results that is different for stressed and nonstressed bottles. The patterns may be compared with model patterns of stressed bottles.

8.4. Voice

Machine intelligence is a very difficult thing to assess and quantify. It is also very subjective in that the "level of intellect" desired varies from one application to another. What is certain is that once a move away from the rudimentary two-state response is desired, then the use of microelectronics becomes absolutely necessary. This is because of the need for low-cost, high-power modules that can be integrated to give human-compatible communication of information.

The sense that offers a lot of potential in the near future is that of auditory communication, whereby man and machine communicate in the human's natural language. Today's voice recognition devices understand only a fixed number of individual key words or phrases. Also, most systems require that each user speak those words to the computer in advance of their "automatic" recognition. This is to train the machine or to familiarize the machine with the nuances of the speaker's voice. Each key vocabulary word is spoken several times by the user so that the machine can store the digital voice patterns against which it will compare future speech inputs. Thus, when it finds a match it recognizes the word.

At Automan '81 in Brighton, the Unimation Company demonstrated a Unimate Puma robot that was instructed to do tasks by operator speech. While it worked immediately sometimes, at other times the device was somewhat "deaf" to the operator who had to repeat the instructions several times and enunciate slowly.

In the future, voice interaction with computers and machines will be commonplace—as much a part of the computer world as CRT terminals and keyboards are today. Despite major advances in reliability, reducing costs, and an increased naturalness and complexity of recognized utterances, there will still be some restriction as to what can be said by the operator and in what environment. Machines will recognize phrases as subunits of clauses and sentences. Consequently, languages will become more "habitable" so that the "allowable" sentences will be used without the human operator feeling a severe unnaturalness. Nevertheless, it will not be possible in the near future to carry out the unrestricted conversations that occur in normal human intercourse.[31]

The prospect of a limited vocabulary for communication in high technology should not be considered a restriction on its adoption since it already exists and functions within air traffic control. Regular and special words are used so that ambiguity is avoided, yet there are a sufficient number of words that permit a "full" conversation to be conducted by persons "fluent" in the language. Most languages have a multitude of redundant words and synonyms that permit variation and nuances to be used in communication, where preciseness is not paramount. This is the case, for instance, in poetry, novels, normal literature and conversation, yet as soon as preciseness is required, as in say law, then the language becomes very restricted and unique.

An indication of the complexity of speech recognition systems is given by an extract from the BRSL study[2] that deals with pattern and voice recognition. The overview deals with "Hearsay-2," a voice recognition system developed by Carnegie-Mellon University in the United States. In addition to its advancements of speech processing techniques, Hearsay-2 embodies novel concepts that offer a significant contribution to artifical intelligence research (see also section 8.5).

Two of the major concerns of speech recognition are how to represent the knowledge about a particular problem and how to organize the knowledge in such a manner that it may be brought to bear upon the problem. The knowledge that could be potentially applied is diverse and exists at several levels. At the lowest level, such knowledge includes how "phones" are related to raw speech input data. Intermediate level knowledge includes how phones are related to syllables and words. At the highest level, useful knowledge includes how likely a particular word is in a particular point in the utterance, given the words in the other parts of the utterance. This likelihood can be calculated from simple measures such as the probability of one word following another based on tables of word pair frequencies, or from more complex measures utilizing syntactic and semantic analysis of the partially decoded utterance.

A system for speech understanding should be able to use some or all of these areas of knowledge. Hearsay-2 provides a framework within which relevant knowledge can be stored and used to analyze the speech utterance. The knowledge must be divided into small units called Knowledge Sources (KSs) with each KS dealing with a certain aspect of the analysis. A KS is an independent "expert" and does not assume anything about the existence or nature of any of the other KSs. Thus, a KS can be added or removed without having to alter any of the other KSs in the system. There are two advantages to this approach, the first is that new methods of analyzing the utterance may be added with a minimum of effort (only the new KS must be coded), which allows for the testing of different algorithms and combinations of algorithms to be performed easily. This is an important consideration in a field as complex as speech understanding where much development of suitable techniques still has to be done. The second advantage is that the "pool" of KSs may be added to by persons working independently, since detailed knowledge of what other KSs are doing is not required in order to write a new KS.

KSs receive information from, and send information to, other KSs by communicating through a global database called the "blackboard." Information is never transmitted privately from one KS to another. Information on the blackboard is available to all KSs and all data used by all the KSs is stored in the blackboard. The blackboard data may be divided into three classes: (i) raw input, (ii) private data, and (iii) hypotheses. The raw data consists of the lowest level information that is to be analyzed. In the case of speech recognition, this is a numerical coding of the speech waveform. Private data comprises any information that a KS needs to retain between successive calls of the KS. It may also include special messages to be passed from one KS to another, although this practice is not recommended since special knowledge about the KS that is receiving the message is required. This then reduces the independent modularity of the KS.

Hypotheses are the most important data items on the blackboard. It is by the use of hypotheses that the KSs build up an interpretation of the speech input. The final hypothesis is the system's conclusion as to the meaning of the utterance. The guiding principle behind the system's operation is an hypothesis-and-test loop. In essence, a KS will generate a hypothesis about the nature of part or all of the data. Other KSs will test the hypothesis. If it passes the test, the KS is informed, the hypothesis is removed from the blackboard, and, if possible, another hypothesis is generated. As the system contains many KSs, this hypothesis-and-test loop is complicated. A hypothesis made by one KS may cause another KS to make a separate hypothesis, which must itself be tested. In order to deal with this, links between hypotheses are placed on the blackboard generating the KS. The links indicate that one hypothesis supports another and the links are used in two ways. First, a numerical assessment of the correctness of each hypothesis is maintained. The assessment will be higher if there is support from another hypothesis. This assessment is used to determine which hypothesis to work on and whether a conclusion has been reached as to the nature of the utterance. Second, the support links are used to delete hypotheses if a supporting hypothesis is rejected.

The blackboard is divided into levels corresponding to the level of the representation of the parts of the utterance. At the lowest level is the raw speech input data, at the highest level is the sentence that the system hypothesizes the utterance to be. In between, there are various levels of representations from the low to high level: segments, phones, phonemes, syllables, words, and phrases. In addition, the blackboard is divided along a second axis, this time relative to the start of the utterance. Thus, a typical hypothesis might be:

Time 100 msec, word level, hypothesis = "THE"

The blackboard can therefore be considered to be a 2D table accessed via time and level. However, there is an additional structure within the blackboard consisting of the hypothesis support links. These generate a tree structure that extends over different levels and different times.

Each KS will take input from one or more levels and output to one level. The input levels may or may not be the same as the output level. This allows considerable freedom on the flow of information between the levels of representation. A bottom-up approach may be implemented by the KSs passing up information from the lower to the higher levels. For example, raw sensory input is transformed into phonemes, then into words, and finally into a sentence. A top-down approach may be implemented by KSs passing down information from a higher level to a lower level. For example, a possible sentence may be generated by a grammer KS, this may be transformed into phonemes, which are matched into the raw speech data. With the hypothesis-and-test loop, if the phonemes do not match the raw data, the original hypothesis (the generated sentence) is deleted and another tried. In practice, Hearsay-2 uses a combination of top-down and bottom-up approaches.

In this manner the raw speech data is used by KSs to generate a hypothesis about the likely phonemes in some parts of the utterance, which is used to generate plausible words in a bottom-up fashion. Then, high-level KSs generate possible words to fit in the unanalyzed parts of the utterance and these are used to generate possible phonemes in a top-down fashion. A KS operating at the phoneme level attempts to match the phonemes hypothesized from the high level with the phonemes hypothesized from the low level. An important aspect of Hearsay-2 is that it can operate from top-down or bottom-up, or a combination of both, depending on how the KSs are written.

One of the major problems with this type of system is how to activate a KS. Conceptually, each KS is waiting for a particular type of hypothesis to be placed onto the blackboard. When this happens, the KS is activated and the code for the KS is executed. In theory, KSs operate in parallel, since more than one KS may be activated by the addition of a hypothesis to the blackboard. In addition, a KS may have multiple activations, and in Hearsay-2, up to 200 activations may be running at the same time. In practice, the activations must be scheduled. Considerable effort has gone into the development of an efficient scheduling system.

The appearance of speech systems that are proven and ruggedized for industrial operation is some time away. Until the late 1980s, the majority of speech recognizers will be speaker dependent. Initially, the currently dominant isolated word recognition will be progressively replaced by connected word sequences from a small vocabulary. In the early 1990s the systems will be able to handle phrases within continuous spoken sentences of a natural language, in a normal speaker or multiple-speaker format. By the early part of the 21st century, advanced systems will be capable of handling speaking errors, such as false starts, mispronounciations, and several forms of ellipsis in natural discourse.[31]

8.5. Artificial Intelligence

The development of industrial machines with certain senses, especially robots, is causing a gray zone between two previously distinct scientific disciplines, those of machine tools and of artificial intelligence. In the long term it is expected that robots will be able to recognize objects even if they have not seen them before.

The computers in these advanced robots will use "deduction" to work out the meaning of shapes, in the same way that the human brain reaches conclusions about objects that are new to it. It is anticipated that these systems could be installed and be working in industry by the end of this decade.

The world therefore stands on the threshold of a second computer age, with devices that can mimic human thought moving out of the laboratories and into commercial use. Artificial intelligence (AI) is presently being used in three configurations: (i) computerized consultants, (ii) fluent computers, and (iii) artificial senses.[32]

Through the use of "expert programs," AI can be used as intelligent assistants for providing advice and making judgements in specialized areas of expertise. To develop an expert system, researchers often spend years picking the brains of human experts, trying to extract and then structure the elusive knowledge that is the basis of that expert's skill. This task is known as "knowledge engineering," since the more bits of knowledge that can be transferred into the computer, the better the system. To make smarter programs it is believed to be necessary to install in the computer a "fuzziness" capability, whereby it relies less on binary decision making—it is not necessary for something to be black or white before dealing with it—because, with binary decisions, wrong means 100% wrong. On the other hand, if there is a spectrum of situations, then the process is more forgiving.

Fluent computers are systems that understand common parlance immediately and are making computers accessible to anybody that can write. There is no need for humans to learn how to structure questions or commands in computer syntax.

Through the use of AI technology, a computer can rapidly sort through signals coming from cameras and other sources in order to identify images and sounds. It reacts instantly, without the delays that conventional systems require in order to decipher the meaning of a spoken word/sentence or visual image.

In the FAS of the near future, AI programs will drive smart machines and robots, as well as helping plant managers to optimize scheduling and run the entire show. Such plants—the so-called "parts on demand" facilities—will turn out different products on different days, or even hours. Researchers are also exploring the concept of "computerized foremen," who will minimize inventory and production bottlenecks in the FAS. Further, if orders or materials change or if machines breakdown, the systems will automatically change the schedule. It is expected that these AI systems will cut inventory by 5% and improve productivity by 20%. It is also claimed that without AI, it is virtually impossible to optimize more than two machines at once.

Failure Analysis

The fundamental theme of this book is that for the greatest productivity to be realized, there must be quantitative information about all of the elements that constitute the FAS. This chapter indicates how the reliability of the proposed system can be determined (by calculation of the probability of catastrophic failure of the various electromechanical-control systems) through elemental evaluation of the total system.

This elemental evaluation allows for the determination of the various "mean lives" and the corresponding time frame reliabilities. Additionally, redundancy is examined as a means of extending the "mean life" of a subset of elements. The assumption of Poisson frequency of failure is made so that the expected cost due to failure within a specific time frame can be assessed. Likewise, the usage of queuing techniques enables manpower requirements to be predicted so that in-service repair costs are minimized. System failure also refers to the inefficiency of production due to operator variance, rejected work, bad information systems, etc. The application of real time or best guess factors for these elements enables their influence on the total system productivity to be determined.

9.1. Definitions

9.1.1. Reliability

Reliability is the probability that a system will perform satisfactorily for at least a given period of time, when used under stated conditions.

9.1.2. Mission Reliability

Mission reliability is the probability that the system will operate in the mode for which it was designed, given that it was operating in this mode at the beginning of the mission.

9.1.3. Catastrophic Failure

Catastrophic failure occurs when a component within a system becomes completely inoperative or exhibits a gross change in characteristics. While the design of a piece of equipment can sometimes be responsible for such failures, it is usual for them to be considered spontaneous or random in nature.

9.1.4. Out of Tolerance Failures

Out of tolerance failures result from degradation, drift, and wear out. These changes come about as a result of time and environment. When these gradual changes—considered collectively—reach the point when system performance is below acceptable limits, it is said that the system has failed.

9.1.5. Mean Time To Failure (MTTF or θ)

Mean time to failure is the arithmetic average of the life times of the components tested. It is meaningful only in relationship to the frequency distribution assumed by the data. The two usual frequency distributions are:

1. exponential, where the mean life occurs at the point where there is 36.8% probability of survival (see Figure 53);
2. normal, where the mean life occurs at the point where there is 50% probability of survival. The normal frequency distribution differs from the exponential in that it requires both the mean life and the standard deviation before it can be calculated (see Figure 54).

9.1.6. Repairability

Repairability is the probability that a failed system will be restored to operable condition in a specified active repair time.

9.1.7. Active Repair Time

Active repair time is the portion of down time in which personnel are affecting a repair.

9.1.8. Down Time

Down time is the total time during which the system is not in an acceptable operating condition.

9.1.9. Maintainability

Maintainability is the probability that a failed system will be restored to operable condition in a specified down time.

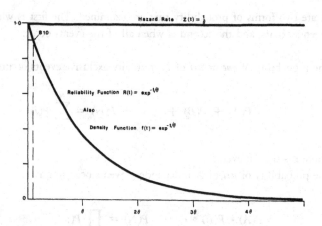

Figure 53. The density function, reliability function, and hazard rate for an exponential frequency distribution.

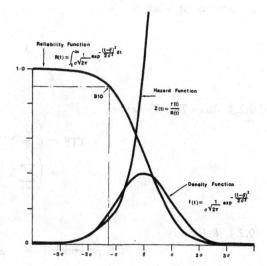

Figure 54. The density function, reliability function, and hazard rate for a normal frequency distribution.

9.2. Mathematical Concepts

9.2.1. Probability

There are two forms of probability that are examined. The first is when one or all of the events occurs, and the second is when all of the events occur:

1. The probability of *one* or *all* of N mutually exclusive events occurring is:

$$P(A) + P(B) + \cdots + P(N) = \sum_{i=A}^{N} P(i)$$

  where i is the ith event.
2. The probability of *all* of N independent events occurring is:

$$P(A) * P(B) * \cdots * P(N) = \prod_{i=A}^{N} P(i)$$

9.2.2. Failure Rate

The failure rate of a component is given as λ occurrences per unit time. For n components the failure rate is:

$$= \sum_{i=1}^{n} \lambda_i$$

9.2.3. Mean Time To Failure

$$\text{MTTF} = \theta = \lambda^{-1}$$
$$= \left(\sum_{i=1}^{n} \lambda_i \right)^{-1}$$

9.2.4. Reliability

$$\text{Exponential: } R = \exp^{-t/\theta}$$

$$\text{Normal: } R(z) = \frac{\theta - t}{\sigma}$$

where exp is the exponential function, σ is the standard deviation, t is the mission time, and $R(z)$ is the area under the right-hand side of a normal curve.

Reliability of a Serial System

$$R_s = \sum_{i=1}^{n} R(i)$$

where R_s is the reliability of the serial system, and i is the ith system.

9.2.5. Frequency of Failure

In multielement systems, the probability of a group of elements of identical failure rate is described by a Poisson distribution:

$$P(k) = \frac{\exp^{-np} * (np)^k}{k!}$$

where p is the probability of one component failing per unit time, n is the number of components, and k is the number of failures.

The Poisson distribution differs from both the exponential and the normal distributions in that it is a discrete function. Hence, it allows for the determination of the probability of n events occurring. A graph of Poisson probabilities for np from 0.1 to 3.0 is given in Figure 55. It can be seen that up to $np = 1.0$, the probability drops as n increases, but with $np > 1.0$, the probability distribution approaches that of the normal distribution as n increases. Also, the probability of > 1 events is greater than and less then that of a single event.

9.3. Redundancy

Redundancy is defined as the existence of more than one means of accomplishing a given task. Thus, in general, all means must fail before there is system failure. Redundancy occurs in a number of different configurations[33]:

- Parallel series: system duplication
- Mixed parallel: unit duplication
- Series parallel: element duplication

Of these, it can be shown that the series-parallel is the most effective. The configuration is available in two forms, that of the "standby parallel" and the "active parallel,"

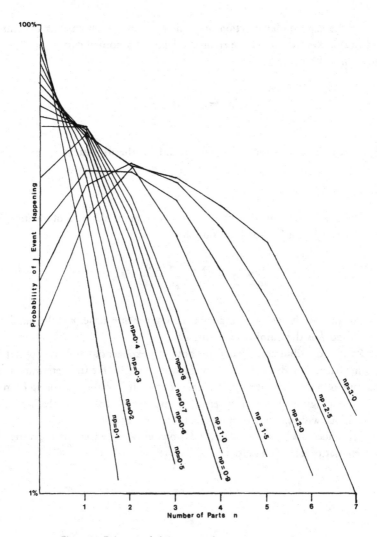

Figure 55. Poisson probability graph for various values of *np*.

where the former represents a dual system with only one unit being activated at a given time, and the latter represents a system with at least two activated units with system reliability being achieved through lowered throughput of each unit or the requirement that only one unit need work for the system to function.

The reliability and mean life formulas for these two systems (assuming two identical elements only) are:

Standby series parallel

$$R(t) = \exp^{-\lambda t}(1 + \lambda t)$$

$$\theta = \sum_{i=1}^{n} \frac{1}{\lambda_i}$$

$$\theta = \frac{2}{\lambda}$$

Active series parallel

$$R(t) = 2 \exp^{-t/\theta} - \exp^{-2t/\theta}$$

$$\theta = \frac{1}{\lambda} \sum_{i=1}^{n} \frac{1}{i}$$

$$\theta = \frac{3}{2\lambda}$$

9.4. Maintenance Considerations

This is the determination of the most cost effective system of maintenance of the system in both its subsystems and totality. Where a nonredundant system exists and where the failure rate is exponential because the hazard rate is constant, it is *not* cost effective to introduce any preventive maintenance procedures. Where a system has redundancy and where the failure rate under consideration is other than exponential, then the introduction of a preventive maintenance program is beneficial.

9.5. Failure Types

In terms of "failure," a component's life is in three stages. When its age-specific failure rate is plotted against time, the graph assumes a bathtub curve as shown in Figure 56. This bathtub curve is an expression of what is generally held to be true about the variation of the instantaneous failure rate of the life cycle of a batch of like components or equipment. In this context, life cycle means the time between the completion of manufacture of an item and it either being scrapped or being overhauled in such a way as to restore the item to "as new" condition. It does not refer to the life cycle of the design. It is also accepted as being a true representation of the variation of the instantaneous failure rate, i.e., the conditional probability of failure or hazard rate for a single component. There are three stages of failure:

1. *Early failures*, which are associated with the running-in period and are due to either substandard parts or excessive stress due to initial maladjustment. Also, they can be due to bad design or the failure of the manufacturer to use adequate quality control.

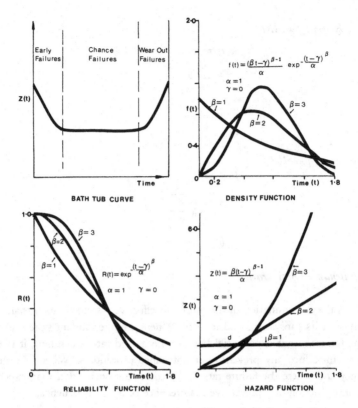

Figure 56. The top left-hand illustration shows the "bathtub" curve that indicates the three stages of failure. The other three illustrations show the density function, reliability function, and hazard function curves for the Weibull distribution. The curves are all drawn for $\beta = 1$, 2, and 3; a γ value of 0; and an α value of 1.

2. *Chance failures* are so-called because they arise from a chance combination of circumstances in design and production. Since these failures occur randomly, they will be randomly distributed in time. Over a long period of time, the failure rate will be constant and the effects of these failures, as well as the probability of their occurrence in specific periods of time, can be readily predicted. It is not possible in any rational way to predict the time at which any such individual event will occur or the time between any particular pair of successive events. Thus, preventing such failures by planned maintenance is not possible; breakdown and repair on occurrence is the inevitable method.

3. *Wear out failures* are caused by components that survive chance failures long enough to encounter the increased failure rate associated with wear out. In this instance, the increasing failure rate and its relationship with age can be predicted (given adequate statistical data about like components), and the whole system returned

to time zero, by overhauling the equipment. Thus, in this case, there is scope for preventive maintenance as a prophylactic.

The use of test or historical data enables a typical life profile of a given component under specific conditions to be plotted. Usage of special graph paper, such as Weibull, allows for the direct determination of the failure stage of the component. Consequently, when a component fails, use of the data on the graph allows for initial determination of the cause of failure. An example of the use of the Weibull graph paper is given in Figure 57. Section 9.8 also gives background information pertaining to the Weibull function. It is considered normal practice to indicate the time at which 10% and 50% of the batch of components could be expected to fail. These are the B10 and B50 points, respectively.

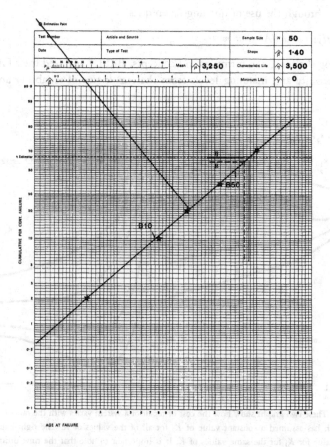

Figure 57. An example of the use of Weibull probability graph paper to interpret data of system failures.

9.6. Manning Levels

The determination of the number of personnel needed to accomplish the maintenance function is dependent upon a number of factors:

(i) *The unit down-time cost,* which is the lost value of production per hour of down time.

(ii) *The cost of maintenance,* which is taken to be the cost to the FAS per man hour of personnel allocated solely to the maintenance of the system.

(iii) *The ratio between inservice failure cost and normal service cost,* for which various curves are drawn in Figure 58. The related costs (not given) indicate that depending upon the ratio, the optimal maintenance schedule varies at time of failure ($k = 1$), to: approximately one third of the MTTF ($k = 10$). Knowledge of the failure rate and the distribution of the repair times for the system being evaluated allow for the minimum cost solutions through the use of queuing techniques.

9.7. Analysis of a FAS System

Using the foregoing formulas and concepts, BRSL[2] analyzed their FAS to determine reliability and manning levels from the standpoint of reliability analysis. The

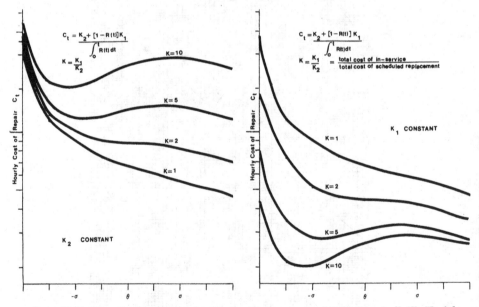

Figure 58. These two graphs show how the cost of repair of a system varies with the ratio K_1/K_2. The left-hand graph has assumed a constant value of K_2 for all of the values of K. The right-hand graph has a constant value for K_1 for the same values of K. It is important to note that the time ordinate should be expressed in hours (days, or years) rather than the statistical values used for these illustrative graphs, if real costs are required.

FAS was analyzed from the "bottom-up". In this manner, a number of elements that could fail were allocated either documented generic failure rates or "educated" failure rates, and the reliability figures for the various subsets were determined. It must be emphasized that because of the dearth of relevant failure data of proprietary components, and since the system was only in its conceptual stage, exponential failure rates for the elements were assumed throughout. It is acknowledged that the "real" system will exhibit a normal failure rate.

The FAS consists of *two* identical production lines, each consisting of *six work areas*. Each work area will (typically) contain *six work stations*, which will be either manual or robotic dependent upon the requisite function of that work station. The ratio of manual to robotic work stations is 21:17; however, depending upon the product being assembled, this ratio could change. Each work area is directly serviced by the stores and there is *no* direct communication between the work areas. Each robotic work station consists (typically) of a triple-headed robot, each robot "head" having up to 7 degrees of freedom. Each work station has an input/output work in progress station on either side of the "operatives" position. The work enters the work areas at a given work station and then progresses to other work stations within that work area, albeit not necessarily unidirectionally.

9.7.1. Analysis of a Nonredundant Robotic Work Area

The analysis assumes that the system is in its steady state condition, that negligible reject work is produced, and that the service levels used for the fully computer-controlled automatic facility are viable, although assumed. The analysis is shown in Figures 59 through 61, with the first figure showing the total facility and the work area–main stores interaction indicated. Figure 60 shows a single-headed robot in elemental form and the calculations that indicate a 90% probability of reliability of operation for an 80-hr two-shift time period. The reliability of a work area consisting of six triple-headed robots and the associated work in progress and conveyoring systems is shown in Figure 61.

BRSL also assumed that when a failure occurred, personnel are immediately available to repair it, and that any components that are required are likewise available. It was further assumed that the mean repair time was 2 hr, and that based on the previous assumption, the active repair time and the down-time are one and the same.

9.7.2. Determination of the Expected Number of Failures

Since reliability is time dependent, it is necessary to specify the mission time. On the basis that the FAS would operate an 80-hr two-shift week, it made sense to use an 80-hr time frame.

9.7.2.1. Expected Number of Robot Failures in Time Period.

$$R(80) = \exp^{-80/761.96}$$
$$= 0.9003 \quad \text{(from Figure 60)}$$

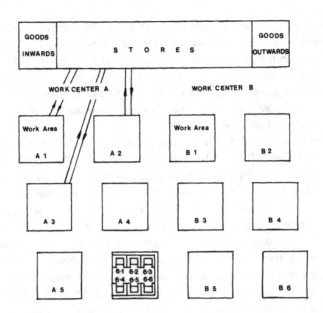

Figure 59. This illustration shows the relationships among the stores, the work areas, and the work stations. Each of the two (identical) work centers has a number of work areas. Within each work area there are a number of work stations (manual and/or robotic). Transactions take place between the stores and work areas within the work centers, they *do not* take place between work areas. Within any one work area, transactions occur between work stations as required for the task to be achieved.

Since there are three robots at each work station, the expected number of robotic failures in the time frame is:

$$E(k)_{80} = \sum_{k=1}^{3} P(k), \qquad P(k) = \frac{\exp^{-np} * (np)^k}{k!}$$

From this we may deduce the probability of one failure as $P(1) = 0.222$, of two failures as $P(2) = 0.033$, of three failures as $P(3) = 0.001$, and so on. Therefore

$$E(k)_{80} = (0.222 * 1) + (0.033 * 2) + (0.001 * 3)$$
$$= 0.291$$

From the above, it can be seen that the probable occurrence of robotic failures is about one every three time periods, or about once every three weeks. It was therefore deemed unnecessary to use redundancy at this level, although there would have to be a sufficient number of spares in stock, to allow for the minimum down time desired for the unit.

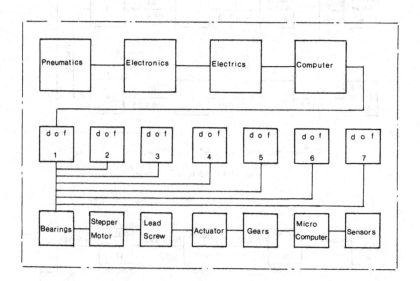

Code	Description	Number in System	Failure Rate(λ) Per Hour x10⁻⁶	Code	Description	Number in System	Failure Rate(λ) Per Hour x10⁻⁶
A	Pneumatic Circuit	1	10	G	Lead Screw 1/dof	7	0·21
B	Electronic Circuit	1	10	H	Actuator 1/dof	7	1
C	Electrical Circuit	1	10	I	Gears 2/dof	14	0·12
D	Computer	1	1	J	Micro Computer 1/dof	7	1
E	Bearing 2/dof	14	0·5	K	Sensors 3/dof	21	3·5
F	Stepper Motor 1/dof	7	0·37				

$$MTTF = \left[\sum_{i=A}^{D} \lambda_i + 7 \sum_{i=F}^{H} \lambda_i + 7\left(2\lambda_E + 2\lambda_I + 3\lambda_K\right) \right]^{-1}$$

$$MTTF = \left\{ \left[10 + 10 + 10 + 1 + 7\left(0·37 + 0·21 + 1\right) + 7\left(2*0·5 + 2*0·12 + 3*3·5\right)\right] *10^{-6} \right\}^{-1}$$

$$MTTF = 761·96 \text{ Hours}$$

$$R_{80} = \exp^{-t/\theta} = \exp^{-80/761·96} = 0·9003$$

Figure 60. This is an idealized analysis of a triple-headed robot in that a lot of the failure rates are educated guesses. It does however give an indicative reliability value for use in later calculations.

9.7.2.2. *Expected Number of Work Station Failures in Time Period.* The MTTF for this unit is less than that of the robot, because the work station incorporates work in progress (wip) units. These wip units are serviced either from the main stores or

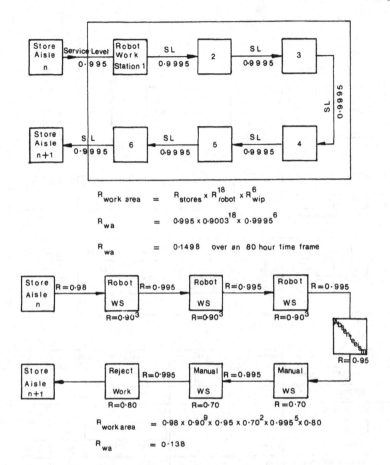

$$R_{work\ area} = R_{stores} \times R_{robot}^{18} \times R_{wip}^{6}$$

$$R_{wa} = 0.995 \times 0.9003^{18} \times 0.9995^{6}$$

$$R_{wa} = 0.1498 \quad \text{over an 80 hour time frame}$$

$$R_{work\ area} = 0.98 \times 0.90^{9} \times 0.95 \times 0.70^{2} \times 0.995^{5} \times 0.80$$

$$R_{wa} = 0.138$$

Figure 61. The reliability of a robotic work area depends upon the reliabilities of the individual modules that constitute that work area. The upper block diagram shows the reliability values for the work stations, the stores and the interwork-station transfer mechanisms. The reliability for the entire work area is given below the block diagram.

The lower block diagram shows the reliability for a mixed robot and manual work area. In this example, values for the reject work and absenteeism (of the work area personnel) are included. The reliability of this particular mix is given below the block diagram.

from the previous work station involved in the assembly task. While the wip systems will have differing service levels based upon their situation, it was assumed (for this calculation) that the lowest service level would be 0.995. This means that the failure rate of the wip unit is 0.005 or that there would be one failure in 200 operations. Therefore, on the assumption of a service call every 0.75 hr, the probable failure rate is once every 150 hr. This means that the wip unit will fail to service the robot(s). It does not necessarily mean that the unit has broken down, only that it cannot service the robot.

$$R_{(\text{work station})} = R_{(\text{robot})}^3 * R_{(\text{wip})}$$
$$= 0.9003^3 * 0.995$$
$$R_{(\text{work station})} = 0.72608 \text{ (over an 80-hr time frame)}$$
$$\lambda_{(\text{work station})} = \frac{\log_e R(t)}{2T}$$
$$= 0.002$$
$$\theta_{(\text{work station})} = 500 \text{ hr}$$

Since there are six work stations in a work area, assuming no redundancies, the expected number of work station failures in the time frame is:

$$E(k)_{80} = \sum_{k=1}^{6} P(k)K \quad \text{where} \quad P(k) = \frac{\exp^{-np} * (np)^k}{k!}$$
$$p = (1 - 0.72608)$$
$$k = 6$$

Therefore

$$E(k)_{80} = 1.6322$$

(which means that an investigation into redundant systems is worthwhile).

9.7.3. Redundancy of Work Stations

Redundant systems are only viable if they meet one of two conditions. The first is if the redundant item if necessary for the safety of the project, i.e., spare engine on an aircraft. The second condition is if the additional cost for the redundant system is less than the anticipated accrued savings. In the case of the FAS, the safety aspects are not applicable; hence, it is purely an economic consideration.

As discussed in Section 9.3, there are two options of redundancy—standby and active. The reliability of a *single* parallel redundant system is determined below:

Standby	Active
$R_r = \exp^{-t/\theta} [1 + (t/\theta)]$	$R_r = 2 \exp^{-t/\theta} - \exp^{-2t/\theta}$
$= \exp^{-80/500} [1 + (80/500)]$	$= 2 \exp^{-80/500} - \exp^{-2(80)/500}$
$R_r = 0.9885$	$R_r = 0.9781$
$\lambda_r = \dfrac{\log_e R(t)}{2T}$	$\lambda_r = \dfrac{\log_e R(t)}{2T}$
$\lambda_r = 0.00007$	$\lambda_r = 0.00014$
$\theta_r = 13{,}816 \text{ hr}$	$\theta_r = 7{,}238 \text{ hr}$

Since in both cases, the *primary* and the *redundant* units have to be capable of operation on their own, the MTTF for each unit is assumed to be the MTTF of the system. Therefore, the expected number of failures in the time frame is:

Expected number of failures in the redundant system, plus expected number of failures in the nonredundant system

$$E(k)_{80} = \sum_{k=1}^{2} P(k)k + \sum_{k=1}^{4} P(k)k$$

Redundant portion of system	Redundant portion of system
$P(k) = \dfrac{\exp^{-np} * (np)^k}{k!}$	$P(k) = \dfrac{\exp^{-np} * (np)^k}{k!}$
where $p = (1 - 0.9885^{1/2})$	where $p = (1 - 0.9781^{1/2})$
$n = 2$	$n = 2$
Therefore	Therefore
$\sum_{k=1}^{2} P(k) = 0.01153$	$\sum_{k=1}^{2} P(k) = 0.02197$

Nonredundant portion of system

$$P(k) = \frac{\exp^{-np} * (np)^k}{k!} \qquad \text{where} \quad p = (1 - 0.7260)$$
$$n = 4$$

$$\sum_{k=1}^{4} P(k) = 1.068$$

Therefore:

$E(k)_{80} = 0.01153 + 1.068$	$E(k)_{80} = 0.02197 + 1.068$
$= 1.07953$	$= 1.0899$

Hence, the use of a single parallel redundant system does not reduce the expected number of work station break downs in a work area, over the prescribed time frame. Other styles of redundancy can be examined to check for viability. This has not been done is this book because of the conceptual nature of the information.

9.7.4. *Manning Levels for In-Service Repairs*

This section is concerned with the determination of the minimum manning level per robotic work area so that in-service repair costs are minimized. The analysis is made on two assumptions:

(i) the expected failure rate;
(ii) the expected service rate to rectify the fault.

It is assumed that a good diagnostic system is an integral part of the work areas and that because of the design of the work station (area), the majority of the faults (i.e., 99.9995%) could be rectified by means of plug-in modules and, consequently, the down time can be minimized. It has also been taken as fact that the maintenance stores will hold modules at a high service level so that the assumed repair times can be realized.

9.7.4.1. *Cost Penalty for System Failure.*

Anticipated value of annual production: $90 million
Hourly cost for lost production: $23,438
Cost per hour penalty for each work center: $11,719

Hence, because each production line (work center) virtually operates in the flow production mode, it follows that failure of *any* work area could constitute a stoppage of the total work center.

9.7.4.2. *Determination of Manning Levels.* Through the use of queuing techniques, the minimal cost manning level is derived by the use of the following model:

$$\text{Poi } (\lambda)/\text{Exp } (1/\mu)/S : \infty, \text{ FIFO}$$

where

Arrival rate $= \lambda/t$ per hour (λ is the expected number of failures in time t)
Service rate $= 1/\mu$ per hour (μ is the mean service time, say two hours)
Number of service personnel $= S$
Length of down time $= L$
Cost of down time $= \$11,719/\text{hr}$
Cost of service $= \$20.00/\text{hr}$
Total cost per hour $= \$20S + \$11,719L$

For a work area without redundancy and for a mean service time of 2 hr, the manning level is calculated thus:

$$\text{Arrivals} = (1.6322/80)/\text{hr}$$
$$= 0.0204$$
$$\text{Services} = (\tfrac{1}{2})/\text{hr}$$
$$\text{Intensity of activity } \rho = \frac{\lambda}{S\mu} = \frac{0.0204}{S * 0.5}$$
$$= \frac{0.0408}{S}$$

$$L = \left(\frac{(S\rho)^s}{S!} * \frac{\rho}{(1-\rho)^2} * P_0 \right) + S\rho$$

$$L = \left\{ \frac{(S\rho)^s}{S!} * \frac{\rho}{(1-\rho)^2} * \left[\left(\sum_{i=0}^{s-1} \frac{(S\rho)^i}{i!} \right) + \frac{(S\rho)^s}{S!(1-\rho)} \right]^{-1} \right\} + S\rho$$

The results for S values of 1, 2, and 3 are given below:

	ρ	P_0	L	$10S$	$11,719L$	TC
1	0.0408	0.9592	0.0425	20	498.06	518.06
2	0.0204	0.9600	0.0408	40	478.48	518.48
3	0.0136	0.9660	0.0408	60	478.13	538.13

It can therefore be seen that, based on a mean service time of 2 hr, one service person per robotic work area is the minimum necessary. This, however, should not be considered to be the only service personnel required, since the queuing formulas take no account of the number of back-up personnel needed to refurbish the failed modules/components.

9.7.5. *Failure Analysis of a Mixed Man–Machine Environment*

All of the previous calculations have been directed towards a totally automated environment and, as such, have enabled an exact failure rate to be determined. Many FASs incorporate manual work stations as a fixed feature *and* as temporary features during the start-up phase. Either way there will be a composite of manual and robotic work stations/areas, and, consequently, the human propensity for a variable, disordered state must be taken into account:

This state is typified by

- nonpredictable and variable performance;
- absenteeism of operatives;
- variable quality of product.

In order to assess this situation, BRSL[2] determined quantitative values for some of the variables:

Nonrefurbishable scrap rate: 3%
Efficiency of inspectors: 98%
Reject work at subassembly level: 10–30%
Operatives attendance level: 95%
Manual station efficiency: 70%
Service level for stores: 98%

A block diagram showing the elements that *could* comprise a work area with mixed human–robotic work stations is given in Figure 61. The assumed mix of three robotic work stations and two manual work stations was taken so that the illustrative reliability calculation could be made.

9.8. The Use of Weibull Probability Paper

While the most acceptable method of determining reliability is testing, it is unfortunately very expensive. Most manufacturers, therefore, rely on methods such as the use of the Weibull distribution function to interpret failure data. The Weibull function permits plotting a straight line graph from real data to project cumulative percentages of failures.

The purpose of the Weibull plot is to estimate, from a sample of failure times, the Weibull constraints. These, in turn, serve to compare actual reliability with a specification or to detect changes from one batch to another. The Weibull function curves are shown in Figure 56.

The three Weibull constraints are:

1. β, the shaping constant, which is the Weibull slope. If $\beta < 1$, then the failures are due to infant mortality. If $\beta = 1$, then the failures are random and the unit is in its useful life period. If $\beta > 1$, then the failure is due to wear out.

2. γ or t_0, the locating constant, is the origin of time, measured on the scale of age at which the failure rate, peculiar to the fitted distribution, begins to operate. If the plot is a straight line it is normal to assume $\gamma = 0$. γ is determined by drawing three equally spaced (absolute measure not probability) horizontal lines and reading the times t_1, t_2, and t_3, corresponding to their intersections with the plot. Hence

$$\gamma = t_0 = \frac{(t_3 * t_1) - (t_2)^2}{(t_3 + t_1) - 2t_2}$$

The calculation of this constant is also necessary if the plot results in a curve, since a straight line can be derived by subtracting its value (and sign) from each failure age.

When using the plot for analysis, γ must be added back to any figures derived from the plot

3. α or μ, the scaling factor allows for determination of the characteristic life, which represents the age at which 63.2% of the failures have occured.

The completed Weibull graph shown in Figure 57 is based on the failure data of 50 units, as given below. From the graph it can be seen that:

$$B10 = 750 \text{ hr}$$
$$B50 = 2800 \text{ hr}$$
$$\text{Mean life} = 3250 \text{ hr}$$
$$\beta = 1.40 \text{ (which implies failure due to wear out,}$$
hence preventive maintenance should be initiated)

Failure Data

Number failing at time	Cumulative failures	Cumulative failure (%)	Time at failure
1	1	2	240
4	5	10	800
5	10	20	1250
9	19	38	2200
16	35	70	4000

Economics of FASs

Any proposal for a FAS has to be backed up by facts. These facts will relate to the anticipated cost of the FAS, its benefit to the company, and the payback period that can be expected from the investment. In order that a convincing argument can be put across, it is necessary that the facts reflect the financial, economic, technical, and social aspects of the FAS. It is also advisable that they are given in a format that is objective as well as being in a manner that does not permit subjective personalities, opinions, or preferences to cloud the issue. Failure to put forward a comprehensive proposal could result in the project being rejected out of hand, irrespective of the amount of technical work that has been done in order to prove its technical viability. Lastly, it is important to remember that the "judges" of the proposal will invariably be of a financial rather than engineering background. Hence, it is imperative to avoid the trap caused by the different thought patterns of accountants and engineers vis à vis what constitutes a viable project. The engineer will have tunnel vision about the technical merits of the proposal, whereas the accountant will have tunnel vision about manpower saving being the one and only index for assessing the project. In the end, it comes down to carefully weighed arguments in both the engineers and accountants languages that will make or break the proposal.

The first fact that must be determined and specified is why an automated system is needed to assemble the product. The proposal *must* be product based, since if there were no product, then there would be no market to be satisfied. Hence, there would not be any (potential) revenue, so why would a FAS or any other system be required? This is not to say that a product must exist as a proven entity before an FAS can be considered, just that the product to be processed must be definable in general technical and marketing terms.

Any product requires a given number of hours to assemble one unit. It also has an annual demand (that is known or projected). If these two factors are combined, the time required to assemble that number of products will result. A year is a finite amount of time and, in terms of the capacity of a FAS, is a function of the number of personnel and/or robots that are available to satisfy the demand for the product. If there are a number of different products, then it is possible that there could be some conflict of resource demand if the FAS tries to satisfy the output demand for all of the products.

If a graph is drawn that shows hours needed per assembly versus annual demand, then a limit can be drawn that equates with the maximum number of products of a given assembly time (per unit) that can be processed by one person. The boundary can be extended to compensate for *n* persons and/or two or three shifts, working. Projection of the assembly time and the annual quantity of a particular product will intersect at a unique point on the graph. If this point is within the boundary, then processing by a manual team is possible. Conversely, the point will indicate the numbers of persons and/or shifts needed to satisfy (just) that product's demand. This will then indicate the potential cost of satisfying the order.

Figure 62 shows a resource graph. As can be seen, a given combination of assembly time and annual demand will either fall within the region enclosed by the ordinates and the boundary or it will be outside the boundary. The graph can also be used to examine the potential manpower saving that could be realized by automating some of the processes. A very crude method of assessing the return for a particular investment is to consider the manpower saving alone:

$$\text{RoI} = \frac{n(\text{annual cost of operator})}{\text{cost of automated alternative}} \times 100$$

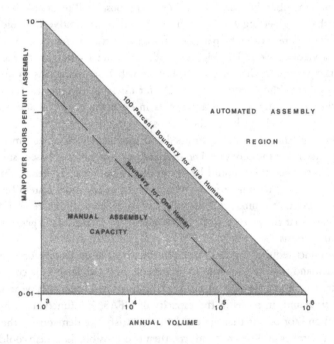

Figure 62. This resource graph shows the capacity of one and five humans for a given range of annual volume of product. The graph boundary can be varied outward for additional humans or inward to take account of the inefficiencies of human productivity. By plotting the annual volume and work content of a particular product it can be determined if the available human resources can accommodate the work load or whether it falls into the region identified as that of automated assembly. This external region merely indicates a larger capacity than that within the boundary. It can therefore be encompassed by increasing the personnel level in the assembly tasks or by using an assembly system that has a larger output per time.

This approximation does not take into account all of the benefits, that can be gotten from an alternative solution in terms of more consistent quality and predictable output. Neither does it take into account depreciation, inflation or the discounted value of money.

To compare the capacity of the FAS against that needed to satisfy the demand for *all* of the products within a given time frame, it is necessary to weigh the various processing times against the total volume being processed. For *n* products, the composite processing time per assembly is:

$$T_n = \sum_{i=1}^{n} \frac{V_i}{V_n} * t_i$$

where T_n is the total hours, V_i is the volume for *i*th product V_n is the total volume of all products, and t_i is the processing time for *i*th product.

It is important to identify what is the correct process rate for the product(s) per unit time, since apart from the pure assembly tasks, there is the time required to set up the work place so that the products can be processed and there is also the time used up in inspection and workplace servicing. The actual value allocated as the boundary should not just be the maximum possible number of hours per *n* man shift year, since there are many factors that can reduce the effective working year.

If manual assembly is considered alone, the processing rate allowed incorporates a factor that takes into account operator fatigue and variation of output. This factor usually increases the time permitted to do a given task by about 20%. In addition, if the operator relies upon automated or semiautomated systems to transfer materials to and from the work place, then there is a high probability that delays will occur. In consequence, the operator may not be able to assemble products. Likewise, illness, natural breaks, union meetings, and other activities reduce the time available for the assembly process. The limiting boundary should therefore be adjusted to reflect the real conditions within the FAS.

If in Figure 62, a product was outside the boundary, then one option would be to use an automated process to assemble the product. Again a resource chart can be drawn as is shown in Figure 63 that indicates the capacity of an automated system. It is known that an automated system has between 1.25 and 2.50 times the potential capacity of a manual system. The capacity is calculated from the number of days available, the number of shifts per day used, and the expected up time for the system. However, this maximum must be reduced by the time required to set up and change over the various tools, the time required for regular maintenance and service, as well as the time needed for any extra programming.

Before a full comparison can be made between one assembly system and an alternate, it is essential that the full and total costs for each system be listed. Quite often, when a proposal is being made up, it is implicitly assumed that the working environment in its *total* form is known deterministically. An indication of some of the tangible and intangible factors involved in the environment are given in Table 20. Exam-

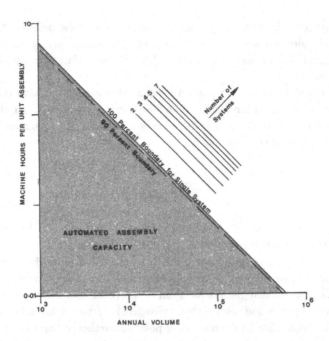

Figure 63. This is a resource graph for an automated system. Here the boundary can be varied by increasing the number of systems and/or decreased by taking into account the down time of the systems or the number of shifts operated if less than three.

Table 20

Factors That Should be Considered When Formulating a Proposal

Tangible factors	Intangible factors
(Mainly technical-economic)	(Mainly socio-organizational)
Cost of machines	Cost of quality control
Cost of jigs and fixtures	Cost of supervision and manpower control
Cost of power supply	Cost of absenteeism
Cost of labor	Cost relative to job satisfaction
Cost of maintenance	Cost due to greivances and strikes
Cost of layout	Cost of organizational flexibility
Cost of setups	Cost of occupational disease

Other factors that have a dramatic effect on the cost of a system are:
- useful life of system
- number of products per unit time
- number of different products, styles, types, ranges, etc.
- time to produce a specific number of products by human
- time to produce a specific number of products by robot
- number of executed operations per product
- life cycle of the product(s)
- flexibility parameters of the system—geometric, functional, and power

ination of these factors will indicate the magnitude of erroneous values that could be applied to a proposal and simultaneously they show that an intelligent application of probability factors to these values will yield a better and more accurate indicator of the expected project costs.

Table 20, by inspection, can also give appreciation of the degree of error that could arise from assuming that all of the values assessed for the factors were absolute. It is therefore necessary to determine which, if any, of these factors are more influencial than the others and which are not deterministically known.

When a factor is not known in quantitative terms (deterministically), it is often known in qualititative terms, whereby it can be said that such and such is liable to happen under a given set of circumstances. An allocation of numerical values to these qualitative terms and conditions is known as stochastic or probabalistic analysis. However, when using any form of statistics, it is wise to take not of the adage:

> Statistics can be used as a drunk uses a lamp post—for support *not* illumination.

Recognition of this danger should lessen the risk of unquestioningly using any data so derived.

10.1. Productivity Indices

Productivity is considered one of the most important, if not the key measurement, of the efficiency of an FAS. Yet there is not a single, unique definition for productivity. The productivity ratios used vary from nation to nation, from industry to industry, and from plant to plant. The measures used are all very subjective and their validity varies according to the industry sector and production class. Sutton,[34] claims that it is necessary to use substitute or surrogate production ratios that relate to key parameters within the sector/class that is of interest. This is very similar to the analysis ratios that are used by accountants to objectively compare a company's performance against others in the same industrial sector. They are also used to compare different divisions within the same company so that corrective actions can be applied where necessary and the results highlighted.

The values and indices listed in Tables 21 and 22 are neither absolute nor exclusive since different assessment criteria may require new or unique ratios to be constructed. What is certain is that the information used for the assessment must be up to date, relevant, and valid, if the ratios are to be of benefit.

10.2. Economics of Alternative Systems

When comparing alternative assembly systems it is important to know what is being compared and against what criteria. For example, it is essential that when comparing operating costs it is for the same quantity or that operating efficiencies are

Table 21
Substitute Productivity Ratios for Automated Production Equipment

$$\text{Production index} = \frac{\text{Recorded production hours}}{\text{Total scheduled hours}}$$

$$\text{Availability index} = \frac{\text{Total scheduled hours} - \text{recorded maintenance hours}}{\text{Total scheduled hours}}$$

$$\text{Utilization index} = \frac{\text{Recorded production hours} + \text{recorded maintenance hours}}{\text{Total hours in time period considered}}$$

$$\text{Cycle cutting ratio} = \frac{\text{Machine cutting time}}{\text{Total cycle time, including set-up, loading, and tool change}}$$

$$\text{Expansion ratio} = \frac{\text{Scheduled production hours}}{\text{Maximum possible production hours available}}$$

$$\text{Maintenance ratio} = \frac{\text{Maintenance cost of machines}}{\text{Machine hours available}}$$

$$\text{Output per hour} = \frac{\text{Number of parts manufactured}}{\text{Machine hours operated}}$$

$$\text{Rejection ratio} = \frac{\text{Number of parts rejected}}{\text{Total number of parts manufactured}}$$

$$\text{Removal index} = \frac{\text{Amount of material removed (weight)}}{\text{Total cutting time}}$$

$$\text{Tooling ratio} = \frac{\text{Cost of tools and fixtures}}{\text{Total production cost}}$$

Table 22
Figures of Merit of Indices Used in Lieu of Productivity Ratios for Evaluating Industry and/or Company Productivity Changes

$$\text{Employment ratio} = \frac{\text{Value added}}{\text{Number of persons employed}}$$

$$\text{Energy ratio} = \frac{\text{Number of units manufactured}}{\text{Energy used (kWh)}}$$

$$\text{Work ratio} = \frac{\text{Employee hours worked}}{\text{Employee hours paid}}$$

$$\text{Material ratio} = \frac{\text{Material in final product}}{\text{Total material supplied}}$$

$$\text{Quality ratio} = \frac{\text{Total cost of quality assurance}}{\text{Total product cost}}$$

$$\text{Cost per part} = \frac{\text{Total operating cost}}{\text{Number of parts produced}}$$

$$\text{Capital intensity} = \frac{\text{Capital equipment employed}}{\text{Number of human employees involed in process}}$$

$$\text{On-time delivery ratio} = \frac{\text{Number of on-time deliveries}}{\text{Total number of deliveries}}$$

$$\text{Inventory turnover} = \frac{\text{Cost of goods sold}}{\text{Average inventory for the period assessed}}$$

$$\text{Recycled material ratio} = \frac{\text{Recycled material used (weight)}}{\text{Total material used (weight)}}$$

compared over the same time period. The determination of an alernative assembly system can be made on the basis of the proposal for automating an existing system without necessarily wanting to increase the production quantity, or on the comparison of two automated systems to produce a given quantity of products or the cost per unit across a range of quantities of a given product.

The determination of one of the three alternatives of manual, flexible automation, or hard automation systems for processing a batch of products is the subject of many research and conference papers. One piece of research uses the product of the system's assembly time and the cost of the system so as to give a value in dollars $\times$ sec. In this way, each alternate can be allocated the appropriate value of the measure and so be compared. The advantage is that while different systems may have different values for either the time or the cost, providing that the price–time product is the same, then it does not matter which of the systems is used. This is on the assumption that one or other of the two factors did not rule a particular system out because of other criteria.[35] For instance, if one alternate cost $100,000 and took 2 sec to assemble the item and the second alternate took 4 seconds *but* cost only $50,000, then they both have a price–time product of 200,000 $ $\times$ sec. The same research has shown that based upon typical values of system components, labor costs, etc., for a flexible system to be economically viable against either manual or hard automation alternatives its price–time product should not exceed $293,000 $ $\times$ sec.

Figures 64 and 65 show the economic relationships for manual, fixed automation, and programmable assembly for a certain set of assumptions. The graph indicates the cost curves for each alternate and the break-even point is given for two of the options. The first is between manual and fixed automatic assembly and the second is between annual, fixed automation, and programmable assembly solutions.[35]

Another assessment procedure uses an equilateral triangle with logarithmic scales on all three sides to determine what the economic advantage between the three alternative methods is. The basic equation of the triangular relationships is[36]:

$$\frac{C}{T} = Mt$$

where C is the cost of assembly, M is the cost of operating an assembly station per unit time, n is the number of parts assembled to the product, and t is the average assembly time per part assembled in the product.

While it seems like a very simple relationship, in reality it does include 22 variables. The equation is also modified into three forms, one for each assembly alternate. The equations are:

(i) Manual assembly:

$$C = kt_M (1 + X)\left(\frac{n}{k} W\right) + \frac{nW}{ksQ}(2C_M + YC_P)$$

Manual assembly cost:

$$MCPU = MATP * LABCST * NPART \longleftarrow$$

where   MCPU = manual assembly cost per unit

MATP = manual assembly time per part, sec

LABCST = cost rate of labor, $/sec

NPART = number of parts in the product

Programmable system cost:

$$PSCST = NSTA * STAP + NPART * TOLPP$$

where   PSCST = programmable system cost

NSTA = number of assembly stations

STAP   = single station price

*TOLPP = tooling price per part

he number of stations required is

$$NSTA = \frac{VOL * NPART * PARTTIME}{NSPY}$$

where   PARTTIME = assembly time per part, sec

NSPY = 1.152×10^7 sec/yr for an uptime fraction of

            0.8, a 250 day year and a 2 shift, 16 hour day.

The assembly system cost is then

$$PSCST = \frac{STAP * VOL * NPART * PARTTIME}{NSPY} + NPART * TOLPP$$

and, using the same payback period model as in eq(11-27), the assembly cost per unit is

$$PCPU = \frac{NPART}{PAYPER} \left[\frac{STAP * PARTTIME}{NSPY} + \frac{TOLPP}{VOL} \right] \longleftarrow$$

---

*Includes basic feeding mechanism (bowl feeders, hoppers, magazines, etc), feed tracks and chutes, and placement or escapement devices or conveying mechanisms that link the parts together.

Fixed automation (transfer machine) assembly cost:

$$TCPU = \frac{TSCST}{PAYPER * VOL}$$

where   TCPU = transfer machine assembly cost per unit

TSCST = transfer machine total cost

PAYPER = payback period in years

Next,   TSCST = NPART * TMCPP

where   TMCPP = transfer machine cost per part

Therefore

$$TCPU = \frac{NPART * TMCPP}{PAYPER * VOL} \longleftarrow$$

This model is based on the payback period method rather than discounted cash flow for simplicity. Any particular case can be worked out using the more accurate method.

Figure 64. These are the economic models for manual, programmable (flexible) automation, and hard automation.

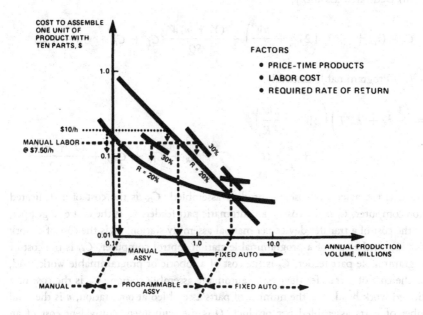

ASSUMPTIONS

NUMBER OF PARTS	=	10
PROGRAMMABLE STATION PRICE	=	$30,000
TOOLING PER PART	=	$7500
TRANSFER MACHINE COST PER PART	=	$30,000
PART STATION TIME	=	3 s
PAYBACK PERIOD	=	2 years to 4 years
LABOR COST	=	$7.50/hr to $10.00/hr
MANUAL ASSEMBLY STATION TIME	=	7 s
NUMBER OF SECONDS PER YEAR	=	1.152×10^7

Figure 65. The top illustration is a graphical representation of the economic models given in Figure 64 based upon the cost assumptions listed in the lower illustration.

(ii) Dedicated assembly:

$$C = (t_D + nXT)\left(2W + \frac{nW}{3k}\right) + \frac{(Y + n)W}{SQ}(C_T + C_P + C_W + C_{DC})$$

(iii) Programmable assembly:

$$C = \left(\frac{k}{2}t_P + kXT\right)\left(2W + \frac{nW}{3K}\right)$$
$$+ \frac{nW}{kSQ}(C_T + C_P + C_R + C_{PF} + C_{PC}) + \frac{(Y + n)W}{SQ}C_G$$

where C is the assembly cost per product assembled, C_{DC} is the cost of a dedicated station computer, C_F is the cost of an automatic part feeder, C_G is the cost of a gripper, C_M is the cost of a transfer device per manual assembly station, C_P is the cost of a work carrier, C_{PC} is the cost of a programmable station control computer, C_{PF} is the cost of a programmable part feeder, C_R is the cost of a robotic or programmable work head, C_T is the cost of a transfer device per automatic assembly station, C_W is the cost of a dedicated work head, k is the number of parts assembled at one station, n is the total number of parts assembled per product, Q is the equipment equivalent cost of an assembly operator, S is the number of shifts per day, T is the machine down time due to defective parts, t_D is the average assembly time per part (dedicated), t_M is the average assembly time per part (manual), t_P is the average assembly time per part (programmable), W is the manual assembly operator labor rate (including overhead), W_M is the total cost of manual assembly (including overheads) $(n/k)W$, X is the ratio of defective to acceptable parts, and Y is the number of product styles (as a function of assembly).

As with the first process, a resource graph is constructed to determine if the product is a candidate for automated assembly. For those products deemed appropriate, the annual volume is plotted on the base line of the triangle and a perpendicular line is drawn. The number of product styles is then plotted on the right-hand side of the triangle and another perpendicular line is drawn so as to intersect with the first. From this intersection, a third line is drawn perpendicular to the left-hand side of the triangle. This last line passes through the three parallel lines that represent the relative cost of assembly per part produced. Where one automated assembly method dominates strongly over the other, only that scale will be "marked." The scale representing manual assembly is there for two purposes: one is to provide a base for converting from relative to specific cost, the other is to provide a reference line so that once the selection has been made between the two automated alternatives, the economics of the selected method to that of the (existing) manual method can be performed.

Figure 66 shows the selection triangle marked up for two test cases. The first (1,2,3) shows that the flexible automated option is best and it would cost approxi-

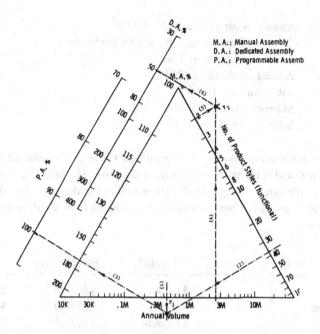

Figure 66. This selection triangle is marked up for two test cases (1,2,3,) and (4,5,6,). The three parallel lines on the left-hand side of the triangle give the relative costs of the three alternate assembly methods. [Courtesy of IFS(Publications)Ltd., Bedford, England.]

mately 0.6 times that of the cost if performed by manual methods. The second case indicates that dedicated automation is the best choice with the assembly cost being 0.5 (50%/99%) times that of the manual process.

A third method of comparing he relative costs of the three process options is that of direct calculation. This is relatively quick for the specific quantity, but is time consuming if a number of conditions are to be assessed. Hence, it pays to use graphical methods for the initial work and mathematical methods when exact and finalized information is needed. The information for the calculated solution is [37]:

Manual system: 4 operators @ $8,000/yr

Single 8-hr shift

Reject level of 5%

1 inspector @ $12,000/yr

Allocated overhead, $2,000

Cycle time, 15 sec

Flexible automation: Cost of system, $80,000

Depreciation, five years straight line

Cycle time, 10 sec

Hard automation: Cost of system, $100,000
                      Depreciation, five years straight line
                      Cycle time, 1 sec
Product: Annual production, 480,000
            Labor content, $0.10
            Material content, $5.90
            Selling price, $10.00

For the initial comparison, it is assumed that the inspection and allocated overhead is required and is the same for all three alternatives. The following tabular information shows the various categories of costs as well as the derived values that permit an initial assessment of the benefits that can be obtained from the three options for a given batch size.

Item	Manual	Flexible auto	Hard auto
System cost	$1,000	$80,000	$200,000
Depreciation	NA	$16,000	$40,000
Fixed costs	$14,000	$30,000	$54,000
Variable costs	$6.00	$5.90	$5.90
Selling price	$10.00	$1.00	$10.00
Break-even point	3,500	7,317	13,170

For a batch of 480,000 units

	Manual	Flexible auto	Hard auto
Production cost	$2,894,000	$2,862,000	$2,886,000
Cost reduction vs. manual	NA	$32,000	$8,000
Payback period	NA	2.50 yr	25 yr
Cost per assembly	$6.029	$5.9624	$6.0124

It can be seen that the flexible automation option offers the lowest per unit cost, with a payback period of two-and-a-half years. What is of interest is to calculate the other cost savings that could occur so that a probable total benefit from adopting this system can be obtained.

The additional benefits will come from the increased consistency of quality that is synonymous with automated assembly. The first of these benefits is the removal or reduction of scrapped materials. Irrespective of the cause of scrapped materials the net result is that there is an often irredeemable loss in terms of material, labor, and energy costs, that have been expended in processing a product that is now deemed to be worthless. It should also be remembered that there is an additional cost in terms of the loss of revenue because the now rejected products cannot be sold.

When a product fails within its warranty period, it is necessary for the manufacturer to replace or refurbish it, with the total cost being borne by the manufacturer. The use of an assembly process that is known and defined in terms of quality of process and components will enable the product's reliability to be determined. Hence, the number of "within-warranty" failures can be predicted and the costing structure adjusted to compensate for this cost.

An external inspection facility is only really necessary if the production system assembles products of varying quality. The products can then be tested and those that are deemed to be unsatisfactory can be rejected and/or returned for rectification. With a process that guarantees products of a known and acceptable standard, inspection as a production task must be challenged. The cost of inspection must relate to the number of inspectors involved, as well as the probability of rejected units being accepted by a client and the resulting cost and embarrassment to the manufacturer for that erroneous acceptance. If the total cost of these rejects exceeds the cost of inspection, then inspection as an external task to the assembly processes *must* be reconsidered.

While an automated system might not be any faster than a manual system, the output per shift of acceptable products is invariably many times better. This means that in order to achieve a given number of products, an automated system can do it without the expense of (supervisory) overtime or multishift working, with its additional costs. It also means that the total capacity of the FAS is increased, so that more revenue is possible for the same time frame.

An automated system means that demand fluctuations can be allowed for by varying the operating time of the system. This is better all around than the manual system of hiring, training, and firing people as the demand for products varies. Finally, with a system that has a predictable output per unit time, there will be a reduction in the number of orders lost through stock out and a saving in the inventory of products as well as components needed to cover for scrap and/or erratic output.

With the proposed flexible assembly system it is expected that there is an 0.80 probability that a 20% reduction in the reject rate will occur. If this is not so, then the reject rate will reduce by only 5%. Therefore:

$$\text{Expected savings} = (0.8 \times 0.2 \times 0.05) + (0.2 \times 0.05 \times 0.05)$$
$$\times 480{,}000 \times \$5.90$$
$$= \$26{,}072$$

It is also anticipated that there will be a 0.50 probability that the increased production resulting from the reduction of rejects can be sold without recourse to extended stocking or reduction of price. Therefore:

$$\text{Expected savings} = \text{expected reduction in rejects} \times 0.50 \times \$4.10$$
$$= \$8{,}364$$

It is anticipated that there is a 0.4 and 0.6 probability that the inspection costs will be reduced by 75% and 50%, respectively. Therefore:

$$\text{Expected savings} = (0.04 \times 0.75) + (0.06 \times 0.50) \times \$12,000$$
$$= \$7,200$$

The anticipated savings from the above nonmanpower activities is $40,436. This is 29½% greater than the savings from manpower alone. The anticipated total savings from this simplified analysis is $73,436, which gives a payback period of 1.09 years.

For a specific number of products, the actual cost per assembly by the manual operator can be calculated, but when this comparison is across a range of quantities, the manual cost is assumed to be fixed at the cost of processing a single unit. This is calculated on the basis of the operator's rate (or salary) and the time to complete that one assembly. This comparison across a wide range of quantities is the only way to compare automated systems, and it is noticable that when the quantities are large, then the fixed costs of the system become an insignificant proportion of the assembly's prime cost. Thus, the cost advantage of hard automation only shows itself when the annual demand is in excess of the capability of the alternative system. Using the data for the three alternative systems (given above) the graph shown in Figure 14 (on p. 44), gives the relationships between them.

10.3. Economic Assessment of the Proposed FAS

The economics of the BRSL-proposed FAS[2] are based upon an annual product demand of 300,000 units that have an average selling price of $300 each, which will yield a turnover of $90,000,000 per annum. The cost of materials within the product, compared with a regular factory, were seen to be capable of being reduced by 10%, reflecting principally a significant reduction in the amount of scrap. Labor costs for the direct staff involved in the assembly of the products was deemed to be an average of $20,000 for each of the 162 personnel.

A total of 28 overhead staff were anticipated as being required at an annual cost of $20,000 each for the work area service and the low technology maintenance personnel, $40,000 per annum for each of the five high-technology maintenance staff, the six industrial engineers, and the two shift managers, together with $80,000 for the FAS director.

Table 23 shows a cost summary comparison between the FAS and a conventional factory based on the same product/output requirements. The cost of sales for the FAS is made up of $32.58 M for materials, $3.24 M for labor, and $4.14 M for warranty and freighting costs. There is also a 5% contingency to cover factors such as power failures, vendor supply problems, etc.

The reduction in the costs for materials/production management is due to three

Table 23
Cost Summary Comparison for the Same Product—Output[a]*

Cost item	Conventional	FAS
Sales	90,000	90,000
Cost of sales (direct)	49,500	41,750
Contribution	40,500	48,250
Overheads		
Factory		
Production and material management		640
Industrial engineering		1,200
Material purchasing, scheduling, and handling	8,100	886
Production overheads		730
Property maintenance and utilities		1,200
Other overhead staff		1,280
Research and development	6,300	6,300
Quality assurance	1,800	1,400
Depreciation	3,600	5,066
Sales and administration		
Marketing		
Service and finance	11,700	11,700
Personnel		
General administration		
Total overheads	31,500	30,402
Operating profit	$9,000	$17,848

[a] NOTE: All values in thousands of dollars.

primary factors. These are the high level of automation in the FAS's stores area, the fact that management of the conversion tasks is becoming more autonomous, and (at a grass roots level) increased job satisfaction with the status change, from shop floor staff to supervisory status for many members of the direct labor force.

The increased depreciation reflects the large amounts of capital equipment that will be used within the FAS. The computers and robots are written off over three years and the warehouse equipment over five years. Likewise, the high utilization of automation and computing techniques is highlighted by the reduced need for quality assurance, because the major emphasis is toward built-in quality and away from the concept of inspected-in quality.

It can be seen that the largest perceived savings will be in the direct labor and materials costs. Yet at the same time, the overhead (inclusive of capital equipment depreciation) has not significantly changed. Hence, the use of an automated solution has resulted in a substantial gearing for the system. Likewise, in the break-even analysis given in Figure 67, it can be seen that the FAS has a lower break-even point than that of the conventional system, hence the profit is considerably higher. This means that the FAS can operate at a lower capacity for the same amount of profit. It also

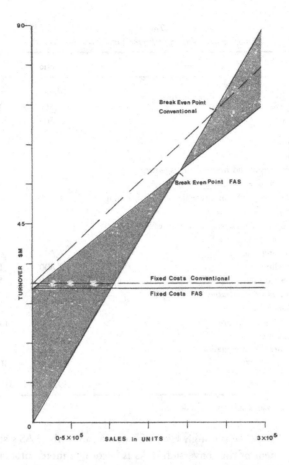

Figure 67. This illustration shows the cost elements for the BRSL-evaluated FAS compared with the costs for the same elements within a conventional factory. As can be seen, both the fixed and variable costs are lower for the FAS, which gives a far lower break-even point with all of the associated benefits.

means that price elasticity can be applied to FAS's products (if required) to compensate for the vagaries of demand for its products. This can be due to reluctance by the consumer to purchase (for any number of reasons that have nothing to do with the FAS), or it could be because of competition from the FAS's rivals who are obtaining a large market share through a price war. Either way, the FAS can afford to slash its selling prices for the goods.

One of the most important factors that must be presented in any economic assessment is that of the bottom line cost of the item being proposed—in this case, an FAS. As can be seen from Table 24, there are four cost areas. The first is the warehouse that totals $4,312,000 and was explained in depth in Chapter 3. The second cost area was

Table 24
Cost Breakdown of the BRSL Proposed FAS

Warehouse	
Pallet and ancillary stores	240,000
Picker cranes	960,000
Racking and bins	1,992,000
Specially molded trays	10,000
Lifts	480,000
Installation and commissioning	590,400
Conveyors	40,000
SUBTOTAL	$4,312,000
Work stations	
Robot	2,063,200
Manual	1,020,400
Conveyors	120,000
PCB line	1,264,000
Engineering and design	368,000
SUBTOTAL	$4,835,600
Computer control	
Hardware	1,000,000
Software	1,000,000
Special equipment	1,000,000
SUBTOTAL	$3,000,000
Land and fabric	$6,000,000
Total likely cost of FAS	$18,147,600

discussed in detail in Chapter 4 and totals $4,835,600 for all of the work stations and associated equipment within the FAS. While the entire FAS is computer controlled in some way or other, the cost of this equipment is only about 16% of the cost of the total FAS. The last cost area is that of the land and fabric of the FAS, which constitutes almost a third of the total cost of $18,147,600.

FAS of the Future

The FAS of today is one of the twin precursors and fundamental elements of the CAF of tomorrow. These CAFs will be totally computerized and integrated so that maximum productivity is assured. In general, the number of machines installed will be less than in today's conventional factories, although they will be more comprehensive and sophisticated. The operating efficiencies of the CAF will enable small-batch production to be as economic as if large-scale production techniques were used. The number of humans involved in the CAF will be small, but they will be large in terms of implied political and social influence. They will also have to be of a high academic background and/or be extremely competent in electronics, mechanics, manufacturing, and computing techniques.

There are two basic opinions as to the exterior appearance of the CAF. The first is that it will be totally, or mainly, built below the ground so as to conserve energy and/or offer a better chance of surviving a nuclear attack. The second opinion is that the CAF will be like the hub of a wheel, with other (mainly) supply plants on the "rim" of this wheel. Transportation routes throughout the CAF complex will form the "spokes" of the wheel, so that there will be a total infrastructure built about and within the CAF, whereby they all exist for each other and none could exist without the others.

Irrespective of the exact external vista the interior of any CAF is defined, except for the specialized machines. The inside will be clinically clean with all of the waste of manufacturing removed for sanitizing and reclamation. The lighting and temperature levels will be consistent with machine requirements in that the processing areas will be dimly lit—since machines do not necessarily use or need visible light to perform their duties—and the temperature, while kept constant, will be lower than that expected today. Again, the reason is that machines are (within limits) indifferent to temperature. It is acknowledged that for certain products, temperature correction may have to be made to the manufacturing dimensions of the components so that the temperature difference between that of the processing area and that in the normal consumer environment does not inhibit the functioning of the product from its declared specifications.

In terms of human accommodation, there will be less space allocated to rest rooms, lunch rooms, parking, etc., primarily because there will be fewer personnel around. However, what is provided will be of a high standard, and will be considered appropriate for the "guardians" of the CAF. The mode of dress will tend to be (clean) white coveralls and the tools will be those of an electronic or computing nature.

The "control room" of the CAF will be occupied by people sitting in front of computer display terminals and consoles offering information on all aspects of the operation of the CAF. These people will be involved with the knowledge-based aspects of the system, while the central computer will control and monitor all other processes down to the last microcomputer controlling the least significant function within the CAF.

A typical scenario that will illustrate the functioning of a CAF is that when an order is received (probably through the central computer, and direct from the client's own computer), it will be assessed and scheduled according to the availability of machines, materials, and the designated priority rating of the order. On the assumption that it is not a repeat order, the requirements will be checked against the CAF's own data base for compatibility with the "composites" of its own group technology families. Through the use of the inhouse CAD/CAM systems, the appropriate processing instruction, tooling, materials, etc., will be identified by code numbers and the order is scheduled.

When appropriate, the order enters the conversion sector of the CAF. Raw materials and components will then be released from the computerized warehouse to arrive at the machines and/or work stations as desired. When the order is complete it will be directed to the computerized goods outward warehouse where it will be crated (if required) and then stored in a coded "pigeon hole" until transported to the client.

The control computer need not be of a parochial nature, but it can be connected to the CAF's clients and suppliers. This means that the computer can automatically order material for its warehouse as stocks vary, and can directly and automatically inform clients when their orders are complete. It could simultaneously transmit the appropriate invoices and, if authorized, the communicating computers could arrange the transfer of credits from one account to another.

There have been many delphic studies into what will constitute the manufacturing industry of the near future. An examination of the study conducted by Battelle[38] shows that both the FAS and the philosophies expounded by this book to be relevant to the development of the CAF as well as the economic survival of the post-industrialized nations of the world.

It is generally considered that by the mid-1980s the use of lasers and electron beams will be increased by 50% over their use at the beginning of the decade; also, that adhesives will replace many of the fasteners used in automobiles and appliances, so as to affect both cost and weight savings. This trend will increase rapidly from the latter half of the decade. (See also Chapter 4).

Automated assembly is expected to become more reliable and practical with manual accomodation not being required for assembly of the more accurately formed components. Fifty percent of the direct labor occupied in small-component assembly and 20% of that in automobile related final assembly tasks is expected to be replaced by programmable automated assembly. Parts storage and retrieval will become integrated with the assembly system. Utilization of automated inspection is expected to increase as computer integrated manufacturing is adopted and as devices, sensors, and technology applicable to automated inspection becomes available. The projected utilization is for 70% of assembly systems to use automatic inspection procedures as a matter of course in the near future.

It is projected that 15% of the assembly system will utilize robotic technology. This will be prompted by the increasing cost of labor and the increasing adoption of microelectronics that will enable better control to be achieved. The development of tactile and visual sensors, as well as automated diagnostics, will enable computerized identification of faults and/or potential faults. This will allow for "precatastrophic" repairs to be conducted.

There will be a trend toward smaller production runs due to the shorter life cycle for various products. This means that long production runs that provide an allowance for incremental adjustments in the process will become rare. Frequent changes in product design and shorter production runs require early optimization of the process, especially as the evolutionary process development cycle is being curtailed. Hence, the use of flexible systems that are proven within their "capability profile," although not for specific products, will become more and more appropirate as the only economic assembly process.

New developments in materials technology are expected to impact the activities of those involved in the FASs and CAFs. There will be a continued trend toward the use of lightweight materials such as aluminum, plastics, and reinforced plastics, as well as the use of composites and honeycombs that give a very high part strength to weight ratio. Also, the adoption of high-strength low-alloy steels that have a two-thirds weight saving on regular steels will result in energy savings through transportation costs.

Three of the drivers of the CAF are personnel hazards, energy conservation, and environmental pollution. The necessity to comply with the numerous OSHA regulations that are becoming more and more specific as time goes by, means that either more and more money has to be spent on protecting people from the hazards and noise within the manufacturing processes *or* the number of people employed must be reduced to a minimum and limited in their direct access to the conversion environment in both time and numbers. Energy shortages and the rising cost of energy dictates that energy intensive processes must be designed to incorporate energy conservation aspects. This means that long established, technically successful processes will have to undergo basic and perhaps drastic modifications or conversion if it can be seen that sufficient energy conservation will be gained from the design change. Further, the use of microelectronics for the operation and control of processes and the environment will optimize the energy consumption as compared with the productivity of the CAF. Likewise, the ubiquitous micro's will be used to monitor and control the wastes, emissions, and effluents of the processes, to ensure that the global environment is not further damaged. As with energy conservation, many of the proven processes will have to be modified to negate any damaging byproducts so that the process becomes environmentally and socially acceptable. Also, new operations will have to have a broad-based environmental analysis supplied with the technical, financial and economic proposals so that the total effect can be assessed.

Another aspect of social ethics, namely, liability, safety, and consumer protection, will be dealt with in a more responsive and rapid manner through the adoption of the CAF than is common nowadays. There have been many cases of product failure, as well as consideration of liability for product failure by courts of law. Many of the failures have been shown to be directly related to human-based activities, which if performed by machines would have yielded a far lower probability of failure occuring and/or not being detected by the process. The necessity to interact with consumer protective agencies and legal institutes requires that a vast amount of information regarding technical activities, maintenance, inspection standards, quality assurance efforts, and other details of corporate and engineering operations will be needed to be kept. The use of a computer-based system means that this information can be directly entered into memory and updated in real time so that when required it can be collated and presented in the desired format.

Ehner and Bax[39] claim that the factory of the future should be examined as functioning in three axes rather than the one-dimensional computer controlled concept that is usually adopted. Each of the three axes can be seen as functioning independently and interdependently with the other two, which allows tradeoff decisions to be made. The three axes are:

X axis: Transformation
Y axis: Finalization
Z axis: Information

The X axis can be described as the maker, in that it includes activities such as material forming and removing, molding, chemical and metallugical processing, painting, and any other activities that control, improve, or replace these processes. The Y axis—finalization—includes all materials handling from manual material movement through robotics to automated guided systems. It also includes assembly and storage; consequently, this axis can be called the mover. The Z axis—information— covers all test and data gathering activities including MRP, production inventory, and cost controls. This axis is seen as the timer in that it sets the speed at which the X axis can pass through the Y axis. It is also responsible for economically minimizing the time between product order and shipment. It is important to note that product engineering and design engineering are not specifically defined within these three parameters. A major portion of those activities will be conducted in the Z axis, albeit in some instances those engineering functions could cover the entire array of axes X, Y, and Z.

Two distinctive characteristics are applicable to this three-axis factory model. First, the axes are not mutually exclusive, since as new technologies are applied, the axes become increasingly interdependent. The classic example is that of the turning process, which has not fundamentally changed (X axis) in many years. However, the use of NC, CNC, and DNC has resulted in Z axis growth that enhances the process's economics, whereas the use of robotic handling equipment in manufacturing cells has caused growth in the Y axis.

The second characteristic of the model is that X, Y, and Z are competitive. Tradeoffs between the three axes, especially between the Y and Z can be made to obtain the desired efficiency and productivity goal. An example of this phenomena is the heavy emphasis that Japan places on flexible automation systems. In this approach, raw materials (X axis) in the bulk form are fed into the parts fabrication process. They are then moved to nonsynchronous assembly lines (Y axis) where at specific points along the line tests and adjustments are made to ensure product quality (Z axis). Thus, stable X axis processes are combined with a large number of Y and relatively few Z activities.

Another way to analyze the characteristics of the three axes is through their effect on the value of the end product. By definition, all X axis processes add to the targeted product value. Only those Y axis processes involved in actual manufacturing (assembly and finishing) add value to the end product. The remaining Y axis and Z axis activities do not add to the product value. These characteristics are demonstrated in the Japanese Kanban (just in time) in-process inventory control system. In this philosophy, the concept of waste includes the activities of setup, maintenance, inventory, material handling and storage, and factory information apart from the normal scrap, rework, and extra cost operations.

In the Kanban system, the goal is to drive all of those activities (that do not add to the product value) as close as possible to zero in order to gain optimal efficiency. Of course, since the activities are interdependent, it is not possible to reach zero. However, the concept does demonstrate that none of the activities add value to the end product and that none is independently more desirable than the others. Thus, the Y and Z axis processes can be regarded as service activities to the X axis functions. The acceptance of this premise permits the recent emphasis on Z axis activities to be seen as part of the transition of the highly industrialized nations of the world moving away from the manufacturing industries and into the service industries. Other manifestations of this transition into a post-industrial society include the evolution of a credit-based economy (rather than one involving cash transactions), the movement to primarily transnational rather than national-based economies, and the fact that many changes are taking place at exponential rather than linear rates.[40]

This three-axis model is an effective means of examining the rate of technological change and determination of the appropriate action for the particular manufacturing application. A separation by axis of the technological areas that have high rates of change is used to form three lists of activities, (as shown in Table 25) which should be monitored as potential candidates for improved factory operations.

The assumption that all common processes in the X axis are stable and exempt

Table 25
Candidates for High Rates of Technological Change

***

X axis processes

Massive fast material removal: continuous creep feed grinding and laser assisted machining.

Minimal material removal: near net shape (NNS) processes, scrap metal extrusion, powder metallurgy, liquid phase sintering, fast plastic molding, and controlled flow casting.

Material treating: new compounds such as amorphous metals.

Metal forming: spin forming and flow turning.

Joining: robotic seam welding, new adhesives, laser and electron beam joining.

Alternative materials: advanced composites with fiberglass, carbon–graphite, boron and its alloys, and ceramics from metal oxides.

Y axis processes

Flexible assembly: nonsynchronous conveyence systems and robots, station skipping, parallel spurs, kit formation and changeover, and instantaneous or very rapid setup changes.

Robotics: grippers, compliance, fixturing, and response time.

Component identification: parts feeders, laser, and vision and position orientation sensors that identify physical and chemical states.

Group/system technology: robot modules, servo devices, product structuring, simulation and optimization, safe shutdown (soft landing).

Z axis processes

Testing: quantification and classification, calibration and feedback, nondestructive testing (NDT), self-diagnostics, and signature analysis.

Sensing: optical, tactile, acoustic, thermal, pressure, chemical, and electrical.

Control: system architecture, algorithimics, computer and actuator compatibility, hardware–software tradeoffs, architecture and manufacturing strategy, database structure, flow and integration, and systems modeling and diagnostics.

***

from technological change (primarily because of the huge investment) should be challenged in the light of current global energy and material supply conditions. These two factors are among the primary drivers of X axis technological change, with material performance, quality, and environment being the other major drivers.

In the Y axis most material handling systems can be eliminated from areas of high growth as can activities that are normally considered to be hard automation. While transfer lines, handling equipment, and other automated Y axis devices will continue to be improved, the main growth areas will come from product proliferation, shortened product design life, and rapid consumer response time. Factors that will drive Y axis technology are inventory control, cycle time, product flexibility, quality, and productivity. Probable areas of high growth include flexible manufacturing assembly systems, industrial robots, and group technology.

It is in the growth of the Z axis where most attention is being placed. This is in part due to the opportunities presented by computer technology. It should be remembered, however, that Z axis activities do not contribute to the product value, they only set the speed at which value is obtained. However, the market's current emphasis on the manufacturer's service response time makes Z axis activities critical to success. This axis consists of both hardware and software components and has many economic drivers such as cash tie-up, data quality, quantity, timeliness, product liability, problem analysis, process optimization, and product proliferation. Areas of particular interest include testing, sensing, communications, interfaces, architecture, and other control sources and system strategies.

The three-axis model can be used to outline the evolution toward the CAF. This evolution is seen in four distinct phases. Technological development in each phase concentrates on a particular axis. Phase 1 is the establishment of manufacturing/ assembly cells where the emphasis is placed on material and inventory movements. The equipment required includes robots that will link the different operations, large transfer lines, and a combination of parts fabrication, assembly, and testing processes. The predominant characteristics will be superior material movement, low inventory, automatic setup between products, fast throughput and cycle times, and automatic scheduling. The high growth areas will be the Y and Z axes, with primary emphasis on the Y axis.

Automated generic manufacturing processes identify phase 2 with new X-axis processes playing the heaviest role. The most probable candidates for change are those activities for which several processes can be combined for maximum productivity. The primary focus will be placed on minimum material use, labor content, processing time and inventory. Developments in these areas will require high growth in X-axis technology, moderate development in Y-axis activities, and an efficient Z-axis process for scheduling and control.

Phase 3 is the integration of the cell/unit information in which the factory timer functions will make the largest contribution. The computer control configuration of each cell and generic manufacturing unit will be linked to a hierachical data processing and control system. Optimal scheduling, minimal inventory backup, minimal processing times between cells/units, and centralized plantwide control will be the standout qualities of this phase.

Unmanned cell/unit operation identifies phase 4 with initially selective unmanned operations in the areas of parts fabrication and assembly. As handling capabilities are increased, the use of unmanned facilities will become more widespread. The emphasis will be on the human and safety controls of equipment, reliable handling, and continuous off-shift operation—incorporating safe shutdown features. These demands call for high reliability in the X-axis processes, optimal use of the new Y-axis technologies, and expanded Z-axis activities that use sensing and correction devices.

The proposed evolution requires full emphasis on all three axes, no single technological base will ensure survival. Improvements in axes will occur in phases to accomplish the complete integration. Individual factories and product lines will have differing time frames and phase progressions, but the evolution into the factories of the future will, at some time, include all of these features.

Conclusion

Throughout this book the emphasis has been on (quantitative) knowledge. The amount of control that can be exercised on the activities within and without a FAS depends directly upon the amount and validity of the knowledge that is available. Too little knowledge means that control is reactive and consequently of a firefighting nature. Too much knowledge is an impossibility, since the FAS is a dynamic entity and therefore the validity and relevance of all knowledge varies with time.

The purpose of a FAS is to process products that are consistent in quality, predictable in output, and reliable in function. The quality and reliability levels are interrelated and reflect the compromise between what is considered the ideal and what is the techno-economic reality of the real world of the FAS. In this respect, the quality level is that required for both the technological satisfaction of the product and the performance of the assembly process. Additionally, the quality level can determine the cosmetic appearance of the product. The economics are the cost per unit of the product, the reliability versus the cost aspects of the purchased materials and components, and the confidence that the FAS has of its suppliers.

The present and future market trends are for products to have a short life cycle, whereby a product's introduction is quickly followed by extinction or where the product goes through the full life cycle of introduction, growth, maturity, and decline for the time of each stage to be contracted. The result is that at maturity, the product either becomes a household name and remains viable for a long time or it suffers very early senility. The means that the manufacturer must have a system that is flexible enough to react to the vagaries of the market forces so that the products supplied are those that are in demand and will be purchased. This is the philosophy of manufacturing against orders and not manufacturing for stock in the hope that the orders will occur. Flexible assembly systems also offer the user a lower break-even point than that which can be expected from conventional assembly systems, hence the user can afford to use price elasticity of the product against the variation of demand.

The FAS as a system is biased toward the total computer control of the stores, processes, and information. The use of computer controlled warehouses, allows high density storage with everything in a known place, and the contents of each place is known to the computer (in real time), in terms of quantity, quality, and identification. The computer controlled warehouse will supply a work station with components or subassemblies for processing, upon demand, based upon a previously agreed schedule. Work stations will generally be of a modular nature that permits easy and effective maintenance to be achieved. Also they can be constructed and reconfigured into many arrangements so as to suit the needs of a particular batch of products in a balanced and efficient subsystem within the total capacity of the FAS. To this end, the effective capacity of the FAS is not limited or measured by the number of work stations used, but by the number of "product dollars" processed in a given time period. It therefore follows that the choice of product is very important, since it is a balance between the value added and the production rate and quantity needed. By optimizing the various factors the productivity of the FAS in terms of revenue can be maximized.

The feature that has the highest profile within the FAS is the robot. These units are extremely varied in configuration, capacity, and intelligence. Their work ethics are totally different from those of people in that they work at a predictable rate, assembling products that are of consistent quality. Depending upon their sophistication, the robots can be as one with their environment, they can be linked with all of the other automated machines and systems to achieve maximum productivity, or they can be very simple units that are used purely for moving material from one place (machine) to another. Intelligence is linked to knowledge, the robots of today can be provided with information through a variety of machine sensors that permit some form of adaptive operation and/or basic decision making to take place. The robot's day of enlightenment will occur by the end of this decade, when artificial intelligence, in its widest sense, is used to provide a total (benevolent) intelligence to support, operate, and sustain the FAS.

There is at present a real need for humans as direct labor for certain assembly operations and for batches of a certain size that precludes the use of either a flexible or hard automation solution. The use of humans within the FAS, working alongside or supervising robots can lead to psycho-sociological problems that will not be resolved until the human accepts the robotic concept of work habits and adopts the correct relationship with advanced automation.

It has been observed that to the world at large, "robots are bad," since they "murder" workers and cause unemployment and havoc throughout industry. This has been shown to be in error, albeit that these assertions are made on the basis of facts which are distorted out of truth.

It is acknowledged that robots, flexible automated systems, and microelectronics will cause, and are causing, many social problems. However, such problems will be short-lived and greatly outweighed by the great social benefits that can be gained from the total acceptance of this megatechnology of microelectronics.

The FAS is a very complex system and consequently is liable to failure at various levels and times. Elemental failure is due to the finite life of the various components and subsystems that constitute the FAS, and while they cannot be prevented from causing disruptions within the FAS, the down time directly attributable to it can be minimized through the use of redundant systems, built-in and integrated automatic diagnostic systems, and modular replacement concepts. This means that the FAS can have an up time of 95%–97%, although this figure will drop significantly if random factors such as humans, incorrectly designed products, poor quality assurance techniques, etc., are allowed to mar the potential of the pure FAS.

The economics of the FAS are a function of the products, components, variable costs, system up time, number of shifts operated, and the percentage capacity that is achieved. As discussed earlier, a FAS has a considerable cushion to absorb the fluctuations in demand for its products. But in the end, the validity of a given FAS is controlled by the revenue derived from sales and the costs incurred by the process.

The adoption rate, sophistication, and development of the FAS will vary over time. At any one point in time, the state of art of the FAS per se is controlled by the interaction of four fundamental forces as shown in Figure 68.

The drivers are those of society and commerce, which are endeavoring to remove humans from the hazardous, boring, and unsocial aspects of the conversion process, and simultaneously to reduce the per unit cost of goods that are created within the manufacturing industry.

The forces of technology and finance are constraining the upward progression of the FAS. However, these are not deliberate obstructing forces, but merely the limits set by the leading edge of today's technology and the cost of technological progress.

Fortunately, the forces are not in balance—otherwise a state of stagnation would occur—but are biased in an upward motion that is increasing as the Western world moves into the status of a postindustrialized society. This means wealth and prosperity to all successful participants and spin-off benefits to the nations that are now entering the manufacturing stage of their evolution.

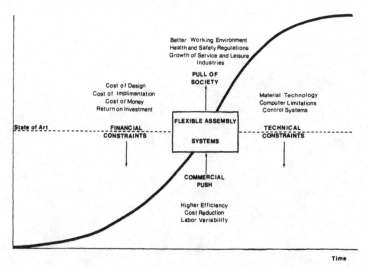

Figure 68. The rate of adoption of the FAS and its inherent philosophies is controlled by four forces. As can be seen, the forces of society and commerce are propelling the acceptance of these systems. The retarding forces are those of technical competence or capability, and financial caution.

The overall scenario is one of a closed-loop global society, with the highly technological countries selling knowledge in exchange for goods manufactured by the developing nations. These latter day manufacturing nations will progress through this stage of their evolution much faster than did their predecessors, so that in the near future all of society will be in the age of enlightenment, where mankind performs all of those tasks that require original thought and creativity, while machines perform the tasks that are necessary and essential for the maintainance of this lifestyle and the expectations of an advanced civilization. To this end FAS is a key.

References

1. M. E. Merchant, World trends in flexible manufacturing systems, *The FMS Magazine* 1(1), 4–5 (1982).
2. British Robotic Systems Ltd., *Final Report on Feasibility Study for U.K. Department of Industry*, Project Number D4N435 (1981). Restricted circulation.
3. J. Hosier, Word processing, *Administration Management* 28(3), 20–21 (1978).
4. F. Jones, Word processing—the Bradford way, conference script, code number AMF0051010/JCM (1979).
5. C. Wyles, Shopfloor troubleshooting by computer, *Engineering Computers* 1(4), 22–25 (1982).
6. M. L. Ernst, The mechanization of commerce, *Scientific American* 247(3), 111–122 (1982).
7. R. Brady, Bar codes, *Management Review & Digest* 8(3), 13–15 (1981).
8. A. Tank, Control all along the line, *Engineering Today*, 6(22), 25–26 (1982).
9. J. Williams, The new warehousing technology, *Management Today: Survey of Physical Distribution*, pp. 23–24 and 26–28, November (1982).
10. A. Scott, Keep materials on the move, *Engineering*, pp. 253–256 and 258–259, April (1982).
11. M. Lee, A car worker's 40% year, *Management Today*, pp. 80–82, February (1980).
12. A. E. (Tony) Owen, The Integrated Factory, Proc. 2nd. Int. Conf. on Assembly Automation, Brighton, U.K., pp. 25–36, May (1981).
13. A. S. Kondoleon, Application of technology-economic model of assembly techniques to programmable assembly machine configurations, Masters thesis, Massachusetts Institute of Technology (1976).
14. A. E. (Tony) Owen, Management of manufacturing industry in the microelectronics age, M.B.A. Dissertation, University of Bradford, U.K. (1979).
15. D. Ward, Where to now R2D2?, *Metalworking Production*, pp. 102–103 and 107, March (1979).
16. H. Voysey, Problems of mingling men and machines, *New Scientist* 75, 416–417, August (1978).
17. J. F. Engelberger, Performance evaluation of industrial robots, *Proc. 3rd. Conf. and 6th Int. Symp. on Industrial Robots*, Nottingham, U.K., Paper J4, March (1976).
18. A. Ioannou and K. Rathmill, Database provides tool for robot selection, *The Industrial Robot* 9(3), 153–157 (1982).
19. B. M. Smith, J. S. Albus, and J. J. Barbera, Draft glossary of terms for robotics, U.S. Department of Commerce, National Bureau of Standards, Washington, D.C. (1979).
20. *Automatic Factory Opportunities in Michigan* (1 Robotics Systems), Industrial Development Division, Institute of Science and Technology, University of Michigan (1982).
21. A. E. (Tony) Owen, *Chips in Industry*, Economist Intelligence Unit, London (1982).
22. I. Asimov, *I, Robot*, Gnome Publishers, New York (1950).
23. N. W. Clapp, Three laws of industrial robotics, *Robotics Today*, pp. 19–22, Spring (1980).
24. W. R. Tanner, Can I use a robot?, *Robotics Today*, pp. 43–44, Spring (1980).
25. The impact of robotics on the workforce and workplace (preliminary draft), Carnegie-Mellon University, (1981).
26. N. Martensson and C. Johansson, Subassembly of a gearshaft by industrial robot, *10th Int. Symp. and 5th Int. Conf. on Industrial Robots*, March (1980), pp. 523–533, Milan, Italy.
27. P. F. Rogers, A time and motion method for industrial robots, *The Industrial Robot* 5(4), 187–192 (1978).

28. A. E. Cullison, Robot 'murders' worker, *The Daily Telegraph,* December 9th 1981.

29. R. J. Barrett, R. Bell, and P. H. Hodson, Planning for robot installation and maintenance: A safety framework, *Proc. 4th British Robot Association Conf.* 13–28, May (1981), Brighton, U.K.

30. C. A. Rosen, in *Computer Vision and Sensor Based Robots* (G. G. Dodd and L. Rossel, eds.), pp. 3–22, Plenum Press, New York (1979).

31. W. A. Lee, Speech recongition, *EDN,* 26(20), 169–170, 172 (1981).

32. Anonymous, Artificial intelligence: the second computer age begins, *Business Week (International Edition),* March 8th, pp. 46–51 (1982).

33. ARINC Research Group, *Reliability Engineering* (W. H. Von Alven, ed.), Prentice-Hall, Princeton (1964).

34. G. P. Sutton, The many faces of productivity, *Manufacturing Engineering* 85(6), 58–64 (1980).

35. J. L. Nevins and D. E. Whitney, in *Computer Vision and Sensor Based Robots* (G. G. Dodd and L. Rossel, eds.) pp. 275–321, Plenum Press, New York (1979).

36. T. Csakvary, Product selection procedure for programmable automatic assembly technique, *Proc. 2nd. Int. Conf. on Automated Assembly,* May 1981 pp. 201–210, Brighton, U.K.

37. A. E. (Tony) Owen, Automated assembly *can* equate with short payback periods, *Proc. 3rd. Int. Conf. on Assembly Automation,* Stuttgart, West Germany, pp. 353–362 (1982).

38. G. W. Levy, Final report on the manufacturing engineer—past, present, and future, Battelle Columbus Laboratories (1979).

39. W. J. Ehner and F. R. Bax, Factory of the future in three axes, *CAD/CAM Technology* 1(2), pp. 26, 29–32 (1982).

40. T. Stonier, The rise and rise of the knowledge industry, *The Business Location File,* June/July, pp. 5–6 (1978).

Bibliography

The following books and articles are given as further reading in alphabetical sequence by author's name. In addition, a list of information sources for the United Kingdom, and the United States is included so that readers can test the pulse of the FAS in real time. Please note that no inference should be drawn from the inclusion or exclusion of books, articles, or information sources in this bibliography.

Acard, *Joining and Assembly: The Impact of Robots and Automation*, Her Majesty's Stationery Office, London, (1979).

B. Atkinson and P. Heywood, *The Challenge of Vision*, British Robotic Systems Ltd. London (1982).

G. Arndt, Integrated flexible manufacturing systems: towards automation in batch production, *New Zealand Engineering* 32(7), 150–155 (1977).

T. Forester, *The Microelectronics Revolution*, Basil Blackwell, Oxford (1980).

G. Friedrichs and A. Schaff, *Microelectronics and Society: For Better or For Worse*, Pergamon Press, Oxford (1982).

Health and Safety Executive (HSE), *Microprocessors in Industry*, Her Majesty's Stationery Office, London (1981).

Ingersoll Engineers, *The FMS Report*, IFS (Publications) Ltd, Kempston (1982).

G. Lundstrom, B. Glemme, and B. W. Rooks, *Industrial Robots–Gripper Review*, IFS (Publications) Ltd, Kempston (1977).

R. Malone, *The Robot Book*, Jove Publications, New York (1978).

G. L. Simons, *Robots in Industry*, NCC, Manchester (1980).

Trades Union Congress, *Employment and Technology*, TC, London (1979).

University of Nottingham, Proceedings of One-Day Seminar on Robot Safety (1982).

A. O. F. Venton, How numerate is terotechnology, *Trans. I. Mar. E.* 87, 202–228 (1975).

For a comprehensive reading list on all aspects of the new technology of microelectronics, see the United Kingdom's Department of Industry's publication, *Information technology—a bibliography*, which is available free from Ashdown House Library, 123 Victoria Street, London SW1E 6RB, England.

Information Sources

Primarily Technical—United Kingdom

Assembly Automation ⎫
FMS Magazine ⎪
Industrial Robot ⎬ IFS (Publications) Ltd., 35–39 High Street,
Sensor Review ⎭ Kempston, Bedford MK42 7BT

CADCAM International

Woodpecker Publications Ltd, 43–45 St. John Street, London EC1M 4AN

Engineer

Morgan–Grampian Publishers Ltd., Calderwood Street, London SE18 6QH

Engineering

The Design Council, 28 Haymarket, London SW1Y 4SU

Engineering Computers

Franks Hall, Horton Kirby, Kent DA4 9LL

Technology

Engineering Today Ltd., 55–57 Great Marlborough Street, London

Primarily Technical—United States

CAD/CAM Technology ⎫
Manufacturing Engineer ⎬
Robotics Today ⎭

One SME Drive, P.O. Box 930, Dearborn, Michigan 48128

Mini–Micro Systems ⎫
EDN ⎬
  ⎭

Cahners Publishing Co., 221 Columbus Avenue, Boston, Massachusetts 02116

Primarily Financial—United Kingdom

None known

Primarily Financial—United States

Computer Integrated Manufacturing (CIM) Newsletter

Prudential–Bache Securities Inc., 100 Gold Street, New York, New York 10292

Weekly Staff Letter

David L. Babson & Co. Inc., One Boston Place, Boston, Massachusetts 02108

Paul Aron Reports

Daiwa Securities America Inc., One Liberty Plaza, New York, New York 10006

Institutes and Academia—United Kingdom

British Robot Association

35–39 High Street, Kempston, Bedford, MK42 7BT

Cranfield Institute of Technology

Cranfield Robotics & Automation Group, Cranfield, Beds MK43 OAL

Production Engineering Research Association (PERA)

Staveley Lodge, Melton Mowbray, Leicester, LE13 OPB

Science & Engineering Research Council (SERC)

Chilton, Didcot, Oxfordshire, OX11 OQX

Institutes and Academia—United States

Carnegie–Mellon University

Department of Engineering and Public Policy, Schenley Park, Pittsburgh, Pennsylvania 15213

Robotics Institute of America ⎫
Society of Manufacturing Engineers ⎬

One SME Drive, P.O. Box 930, Dearborn, Michigan 48128

University of Michigan

Industrial Development Division, 2200 Bonisteel Boulevard, Ann Arbor, Michigan 48109

Audiovisual—United Kingdom

Robots in Industry, Cat. UK2920

Central Film Library, Chalfont Grove, Gerrards Cross, Bucks SL9 8TN

Audiovisual—United States

Industrial Robots—An Introduction

Society of Manufacturing Engineers, One SME Drive, P.O. Box 930, Dearborn, Michigan, 48128

Industrial Robots—Applications

Society of Manufacturing Engineers, One SME Drive, P.O. Box 930, Dearborn, Michigan, 48128

Index